Annals of Mathematics Studies

Number 91

SEMINAR ON SINGULARITIES OF SOLUTIONS OF LINEAR PARTIAL DIFFERENTIAL EQUATIONS

EDITED BY

LARS HÖRMANDER

PRINCETON UNIVERSITY PRESS
AND
UNIVERSITY OF TOKYO PRESS

PRINCETON, NEW JERSEY
1979

Published in Japan exclusively by
University of Tokyo Press;
In other parts of the world by
Princeton University Press

Printed in the United States of America
by Princeton University Press, Princeton, New Jersey

Library of Congress Cataloging in Publication data will
be found on the last printed page of this book

CONTENTS

PREFACE

Apart from some modifications and additions this book consists of the notes of a seminar held at The Institute for Advanced Study in 1977-78. Singularities of solutions of differential equations form a common theme. Some of the lectures were devoted to the analysis of singularities and some to applications in spectral theory.

In an introductory series of lectures the basic techniques of pseudo-differential and Fourier integral operators were developed from this point of view and the basic facts concerning singularities of solutions of operators of principal type were outlined. These lectures form the main part of the first paper here but in the last section some new results on the propagation of singularities (and dual existence theorems) have been added although they were presented later on in the seminar. For operators which are not of principal type but can be written modulo terms of lower order as a product of two such factors, the propagation of singularities is studied in the paper by N. Hanges.

A pseudo-differential operator P is called hypoelliptic if a distribution u can only be singular where Pu is. Two classes of hypoelliptic operators are discussed here. We give a self-contained presentation of Egorov's characterization of subelliptic operators, and the first part of the paper by A. Menikoff is devoted to hypoellipticity of operators with double characteristics.

In the papers mentioned so far the main technical tools have been provided by pseudodifferential operators and the simplest types of Fourier integral operators. However, these are often insufficient to describe the singularities of important objects such as fundamental solutions. Fourier integral operators based on complex phase functions have a much wider scope. They were presented by J. Sjöstrand in two introductory lectures contained in his paper here. The study of boundary problems requires further development of pseudo- and Fourier integral operators. In his paper R. Melrose indicates how one can define these concepts in a manifold with boundary so that the desirable properties are obtained. Applications of these new techniques are outlined in the study of propagation of singularities for mixed boundary problems in the case of non-tangential, diffractive or gliding rays. New results by Melrose and Sjöstrand on propagation of singularities along rays which are tangent to the boundary of finite but arbitrarily high order are given at the end of his paper.

Description of a fundamental solution, say, as a Fourier integral distribution gives a characterization of its singularities. However, it is often of interest to be able to conclude regularity properties of a more classical kind such as smoothness on one side of a hypersurface and existence of a smooth extension across it. The study of lacunas is devoted to such properties, and the paper by L. Boutet de Monvel here sums up some recent work in the area by L. Gårding, A. Hirschowitz and A. Piriou.

Since the pioneering work by Carleman it is well known that there is an intimate relation between the asymptotic properties of eigenfunctions (eigenvalues) of differential operators and the singularities of various transforms of them such as the resolvent, Laplace or Fourier transform. The second part of the paper by Menikoff and that of V. Guillemin are devoted to such applications in spectral theory. For hypoelliptic operators P with double characteristics Menikoff presents joint work with Sjöstrand on the singularities of e^{-tP} as $t \to 0$, and the resulting asymptotic properties of the eigenvalues. The first part of the paper by V. Guillemin is devoted to the remainder term in the asymptotic formulas for the eigenvalues of elliptic operators and the number of lattice points in a large ball. The second part starts from the classical theorem of Szegö on the asymptotic properties of the eigenvalues of the projection of a multiplication operator on spaces of trigonometric polynomials of increasing order. Guillemin substitutes the range of the spectral projections of an elliptic operator for the space of trigonometric polynomials of given degree and obtains similar results with a number of applications. Finally the paper by H. Widom is devoted to the study of higher order terms in the Szegö theorem and the pseudo-differential operator technique which this requires.

This book makes no claim to cover completely the recent work on the singularities of solutions of linear partial differential equations. In particular the analytic (hyperfunction) theory is only mentioned briefly for it was the subject of a parallel seminar. Nevertheless it is hoped that this volume can serve as an introduction to the subject in the absence of a more systematic treatment.

Lars Hörmander

THE INSTITUTE FOR ADVANCED STUDY

Seminar on Singularities
of Solutions of
Linear Partial
Differential Equations

SPECTRAL ANALYSIS OF SINGULARITIES

Lars Hörmander[1)]

1. *Introduction*

The existence of solutions of a linear partial differential equation is closely related to the singularities which solutions of the adjoint equation can have. We shall therefore study singularities of solutions first and discuss existence theorems afterwards as applications.

By the support supp u of a function or distribution u one means the smallest closed set such that u vanishes in the complement. Similarly the singular support sing supp u is the smallest closed set such that u is a C^∞ function in the complement. However, it is possible to make a harmonic analysis of u at the singularities which describes them much more precisely and at the same time simplifies the study of the singularities of solutions of partial differential equations. We shall introduce this harmonic decomposition of the singularities in the next section and then proceed to develop the main tools for studying it in the context of differential equations of principal type where it originated.

2. *Definition and basic properties of the singular spectrum*

If $v \in \mathcal{E}'(\mathbf{R}^n)$ is a distribution with compact support, the Fourier transform $\hat{v}$ is defined by

---

1)Supported in part by NSF grant MCS77-18723.

Seminar on Singularities of Solutions
0-691-08221-9/78/0003 $02.35/1 (cloth)
0-691-08213-8/78/0003 $02.35/1 (paperback)

$$\hat{v}(\xi) = v(e^{-i<\cdot,\xi>}), \; \xi \in \mathbf{R}^n .$$

It is well known that $v \in C^\infty$ if and only if

$$|\hat{v}(\xi)| \leqq C_N(1+|\xi|)^{-N}, \quad N = 1, 2, \cdots ; \; \xi \in \mathbf{R}^n . \tag{2.1}$$

If v is not in C^∞ we can describe the spectrum of the singularities by introducing the cone $\Sigma(v) \subset \mathbf{R}^n \setminus 0$ consisting of directions where (2.1) is not valid. Thus $\eta \notin \Sigma(v)$ means that there is an open cone $V \ni \eta$ such that (2.1) is valid when $\xi \in V$. Clearly $\Sigma(v)$ is a closed cone in $\mathbf{R}^n \setminus 0$, and $\Sigma(v) = \emptyset$ if and only if $v \in C_0^\infty$.

While sing supp v only describes the location of the singularities, the set $\Sigma(v)$ only describes the frequencies causing them. To combine the two types of information we need

LEMMA 2.1. *If* $\phi \in C_0^\infty(\mathbf{R}^n)$ *and* $v \in \mathcal{E}'(\mathbf{R}^n)$, *then*

$$\Sigma(\phi v) \subset \Sigma(v) . \tag{2.2}$$

Proof. The Fourier transform of $u = \phi v$ is $(2\pi)^{-n}\hat{\phi} * \hat{v}$, where $\hat{\phi}$ is rapidly decreasing and $|\hat{v}(\xi)| \leqq C(1+|\xi|)^N$ for some N. If Γ is an open cone where $\hat{v}$ is rapidly decreasing we write

$$|\hat{u}(\xi)| \leqq \int_\Gamma |\hat{\phi}(\xi-\eta)| \, |\hat{v}(\eta)| \, d\eta + C \int_{\complement\Gamma} |\hat{\phi}(\xi-\eta)| \, (1+|\eta|)^N \, d\eta$$

where the integral over Γ is rapidly decreasing as $\xi \to \infty$ because $(1+|\xi|) \leqq (1+|\xi-\eta|)(1+|\eta|)$. This inequality allows us to estimate the second integral by

$$C(1+|\xi|)^N \int_{\complement\Gamma} |\hat{\phi}(\xi-\eta)| \, (1+|\xi-\eta|)^N \, d\eta .$$

If ξ is restricted to a closed cone $\subset \Gamma$, we have $|\xi-\eta| \geqq \varepsilon|\xi|$ for some $\varepsilon > 0$ when $\eta \in \complement\Gamma$, so the integral can be estimated by

$$\int_{|\theta|>\varepsilon|\xi|} |\hat{\phi}(\theta)|\ (1+|\theta|)^N\ d\theta$$

which is rapidly decreasing as $\xi \to \infty$. This proves the lemma.

If $u \in \mathcal{D}'(\Omega)$ where Ω is an open set in $\mathbf{R}^n$, we set for $x \in \Omega$

$$\Sigma_x = \bigcap_\phi \Sigma(\phi u); \quad \phi \in C_0^\infty(\Omega),\ \phi(x) \neq 0\,. \tag{2.3}$$

From (2.2) it follows that

$$\Sigma(\phi u) \to \Sigma_x \ \text{ if } \ \phi \in C_0^\infty(\Omega),\ \phi(x) \neq 0,\ \operatorname{supp} \phi \to \{x\}\,. \tag{2.4}$$

In fact, if V is an open cone $\supset \Sigma_x$, the compactness of the unit sphere shows that one can find $\phi_1, \cdots, \phi_j \in C_0^\infty(\Omega)$ with

$$\phi_1(x)\cdots\phi_j(x) \neq 0\,, \quad \bigcap_1^j \Sigma(\phi_i u) \subset V\,.$$

When $\psi \in C_0^\infty(\Omega)$, $\psi(x) \neq 0$ and $\phi_1 \cdots \phi_j \neq 0$ in $\operatorname{supp} \psi$, we can write $\psi = \phi\phi_1\cdots\phi_j$ where $\phi \in C_0^\infty$, and obtain from (2.2) that

$$\Sigma_x \subset \Sigma(\psi u) \subset \bigcap_1^j \Sigma(\phi_i u) \subset V\,,$$

which proves (2.4).

In particular, (2.4) implies that $\Sigma_x = \emptyset$ if and only if $\phi u \in C^\infty$ for some $\phi \in C_0^\infty(\Omega)$ with $\phi(x) \neq 0$, that is, $x \notin \operatorname{sing\,supp} u$.

DEFINITION 2.2. If $u \in \mathcal{D}'(\Omega)$, then the closed subset of $\Omega \times (\mathbf{R}^n \setminus 0)$ defined by

$$\operatorname{sing\,spec} u = \{(x,\xi) \in \Omega \times (\mathbf{R}^n \setminus 0);\ \xi \in \Sigma_x\}$$

is called the singular spectrum of u.[1] Its projection in Ω is sing supp u.

---

[1] The term wave front set and the notation WF are often used.

Sing spec u is of course conic in the sense that it is invariant under multiplication by positive scalars of the second variable, so it could be considered as a subset of $\Omega \times S^{n-1}$. If $u \in \mathcal{E}'(\mathbf{R}^n)$ it is easily shown by means of a partition of unity that the projection of sing spec u on the second variable is $\Sigma(u)$, so the singular spectrum does contain all information in sing supp u and in $\Sigma(u)$. (However, this projection is not invariant under a change of variables and therefore of limited interest.)

EXAMPLE 2.3. Let $\phi \in C_0^\infty$, $\hat{\phi} \geqq 0$ and $\hat{\phi}(0) = 1$. If $\eta \neq 0$ then

$$u(x) = \sum_1^\infty k^{-2} \phi(kx) e^{ik^2 <x, \eta>}$$

is a continuous function such that sing spec $u = \{(0, t\eta), t > 0\}$. In fact, since every term is in C^∞ and the supports shrink to 0, we have sing supp $u \subset \{0\}$. Now

$$\hat{u}(\xi) = \sum_1^\infty k^{-2-n} \hat{\phi}((\xi - k^2\eta)/k)$$

so $\hat{u}(k^2\eta) \geqq k^{-2-n}$, hence sing spec $u \supset \{(0, t\eta), t>0\}$. If V is an open cone containing η then $|\xi - t\eta| \geqq c(|\xi| + t)$ if $\xi \notin V$ and $t \geqq 0$, for this is true when $|\xi| + t = 1$. Hence $|\xi - k^2\eta|/k \geqq c(|\xi| + k^2)/k \geqq c\sqrt{|\xi|}$. It follows that $\hat{u}$ satisfies (2.1) in $\complement V$ which proves the assertion.

Starting from this example it is easy to show that there is a continuous function u with sing spec u equal to a given conic set. In fact, after suitable topologies have been introduced as we shall do for distributions below, this is a standard condensation of singularities argument.

We shall now discuss how sing spec u transforms if u is composed with a C^∞ map ψ. It is well known that the composition is well defined when the Jacobian matrix ψ' is surjective. However, the notion of singular spectrum allows us to define the composition more generally. To do so we shall first introduce a topology in

$$\mathcal{D}'_\Gamma(\Omega) = \{u \in \mathcal{D}'(\Omega),\ \text{sing spec } u \subset \Gamma\}$$

where Γ is a closed cone $\subset \Omega \times (\mathbf{R}^n \setminus 0)$. Recall that $u \in \mathcal{D}'_\Gamma$ means precisely that for every $\phi \in C_0^\infty(\Omega)$ and every closed cone $V \subset \mathbf{R}^n$ with

(2.5) $$(\text{supp } \phi \times V) \cap \Gamma = \emptyset$$

we have

(2.6) $$\sup_V |\xi|^N |\widehat{\phi u}(\xi)| < \infty, \quad N = 1, 2, \cdots .$$

For a sequence $u_j \in \mathcal{D}'_\Gamma(\Omega)$ we shall say that $u_j \to u$ in $\mathcal{D}'_\Gamma(\Omega)$ if

(i) $$u_j \to u \text{ in } \mathcal{D}'(\Omega) \text{ (weakly)}$$

(ii) $$\sup_V |\xi|^N |\widehat{\phi u_j}(\xi) - \widehat{\phi u}(\xi)| \to 0, \quad j \to \infty,$$

for $N = 1, 2, \cdots$ if $\phi \in C_0^\infty(\Omega)$, V is a closed cone and (2.5) is valid. This implies that $u \in \mathcal{D}'_\Gamma(\Omega)$. Since (i) implies that $\widehat{\phi u_j} \to \widehat{\phi u}$ uniformly on every compact set and N is arbitrary in (ii), we can replace (ii) by

(ii)$'$ $$\sup_j \sup_{\xi \in V} |\xi|^N |\widehat{\phi u_j}(\xi)| < \infty .$$

$C_0^\infty(\Omega)$ is sequentially dense in $\mathcal{D}'_\Gamma(\Omega)$. To prove this we first choose a sequence $\psi_j \in C_0^\infty(\Omega)$ with $0 \leqq \psi_j \leqq 1$ and $\psi_j = 1$ on any compact set for large j, then a sequence $\chi_j \geqq 0$ in $C_0^\infty(\mathbf{R}^n)$ with $\int \chi_j \, dx = 1$ and support shrinking to 0 so fast that $\text{supp } \chi_j + \text{supp } \psi_j \subset \Omega$. Then

$$u_j = \chi_j * (\psi_j u) \in C_0^\infty(\Omega),$$

and $u_j \to u$ in $\mathcal{D}'_\Gamma(\Omega)$. In fact, if $\phi \in C_0^\infty(\Omega)$ we have for large j

$$u_j(\phi) = (\psi_j u)(\check{\chi}_j * \phi) = u(\check{\chi}_j * \phi) \to u(\phi) .$$

If V is a closed cone and (2.5) is valid then we can choose $\psi \in C_0^\infty(\Omega)$ equal to 1 in a neighborhood of $\operatorname{supp} \phi$ and a closed cone W with interior containing $V \setminus 0$ so that $((\operatorname{supp} \psi) \times W) \cap \Gamma = \emptyset$. Hence

$$\phi u_j = \phi(\chi_j * (\psi u)) = \phi w_j$$

for large j, where

$$|\hat{w}_j| = |\hat{\chi}_j|\,|\widehat{\psi u}| \leqq |\widehat{\psi u}|\,.$$

Since $|\widehat{\psi u}|$ is rapidly decreasing in W the proof of Lemma 2.1 gives (ii)$'$ and the assertion is proved.

Theorem 2.4. *Let Ω_j be open subsets of $\mathbf{R}^{n_j}$, $j = 1, 2$, and let $\psi : \Omega_1 \to \Omega_2$ be a C^∞ map. Denote the set of normals of the map by*

$$N_\psi = \{(\psi(y), \xi) \in \Omega_2 \times \mathbf{R}^{n_2}\,;\ {}^t\psi'(y)\xi = 0\}\,.$$

Then the pullback $\psi^* u$ can be defined in one and only one way when $u \in \mathcal{D}'(\Omega_2)$ and $N_\psi \cap$ sing spec $u = \emptyset$ so that it is equal to the composition $u \circ \phi$ when u is a continuous function, and ψ^* is sequentially continuous from $\mathcal{D}'_\Gamma(\Omega_2)$ to $\mathcal{D}'(\Omega_1)$ for any closed cone $\Gamma \subset \Omega_2 \times (\mathbf{R}^{n_2} \setminus 0)$ with $\Gamma \cap N_\psi = \emptyset$. We have

$$\text{sing spec}\, \psi^* u \subset \psi^* \text{sing spec}\, u = \{(y, {}^t\psi'(y)\xi), (\psi(y), \xi) \in \text{sing spec}\, u\}\,.$$

Note that N_ψ consists only of 0 vectors if ψ' is surjective so $\psi^* u$ is always defined then. From the theorem it follows in particular that sing spec u can be defined as a subset of the cotangent bundle $T^*(\Omega)$ with the zero section removed if u is a distribution on the manifold Ω.

The uniqueness statement in Theorem 2.4 is obvious, and for the sufficiency we refer to [7, p. 128-129].

Example 2.5. If u is a C^∞ density on a C^∞ manifold $M \subset \Omega$, then

$$\text{sing spec } u \subset N(M) \setminus 0$$

where $N(M) \subset T^*(\Omega)$ is the conormal bundle; there is equality at points where the density is not 0. In fact, by Theorem 2.4 it suffices to consider the case where M is defined by $x'' = (x_{k+1}, \cdots, x_n) = 0$ and $u = u(x')\,\delta(x'')$ $u \in C^\infty(R^k)$ and $x' = (x_1, \cdots, x_k)$. If u has compact support the Fourier transform $(\xi', \xi'') \to \hat{u}(\xi')$ is rapidly decreasing in all directions with $\xi' \neq 0$ but no direction with $\xi' = 0$, unless $u = 0$, which proves the assertion.

On the other hand, if u is a distribution with $N(M) \cap \text{sing spec } u = \emptyset$ then Theorem 2.4 shows that the restriction of u to M can be defined.

Theorem 2.4 also allows one to extend the definition of multiplication of distributions. First recall that if $u_j \in \mathcal{D}'(\Omega_j)$, $\Omega_j \subset R^{n_j}$, then $u_1 \otimes u_2 \in \mathcal{D}'(\Omega_1 \times \Omega_2)$ is well defined. If $u_j \in \mathcal{E}'$ then the Fourier transform of $u_1 \otimes u_2$ is

$$(\xi_1, \xi_2) \to \hat{u}_1(\xi_1)\hat{u}_2(\xi_2)$$

which is rapidly decreasing except in

$$(\Sigma(u_1) \times \Sigma(u_2)) \cup (0 \times \Sigma(u_2)) \cup (\Sigma(u_1) \times 0)\,.$$

It follows that

$$\text{sing spec } (u_1 \otimes u_2) \subset \{(x_1, x_2; \xi_1, \xi_2);\ (x_j, \xi_j) \in \text{sing spec } u_j \text{ or } \xi_j = 0;\ j = 1, 2\}\,.$$

In particular we can apply this when $\Omega_1 = \Omega_2 = \Omega$. The product $u_1 u_2 \in \mathcal{D}'(\Omega)$ is defined if the pullback of $u_1 \otimes u_2$ by the diagonal map $\Omega \ni x \to (x, x) \in \Omega \times \Omega$ is defined, that is, if $(x, \xi_j) \in \text{sing spec } u_j$ implies $\xi_1 + \xi_2 \neq 0$. In that case

$$\text{sing spec } (u_1 u_2) \subset \{(x, \xi_1 + \xi_2);\ (x, \xi_j) \in \text{sing spec } u_j \text{ or } \xi_j = 0,\ j = 1, 2\}\,.$$

Next we consider the linear transformation defined by a distribution

$K \in \mathcal{D}'(\Omega_1 \times \Omega_2)$ where $\Omega_j \subset R^{n_j}$ are open sets. (The results have obvious extensions to manifolds.) K defines a continuous map $K: C_0^\infty(\Omega_2) \to \mathcal{D}'(\Omega_1)$

$$\langle K\phi, \psi \rangle = K(\psi \otimes \phi);\ \phi \in C_0^\infty(\Omega_2),\ \psi \in C_0^\infty(\Omega_1)\ .$$

Conversely by Schwartz' kernel theorem every such linear map K is defined by a unique distribution K, which justifies our abuse of notations.

THEOREM 2.6. *For any* $u \in C_0^\infty(\Omega_2)$ *we have*

$$\text{sing spec } Ku \subset \text{sing spec}_{\Omega_1} K \underset{(\text{def})}{=} \{(x,\xi);\ (x,\xi,y,0) \in \text{sing spec } K \text{ for some } y \in \Omega_2\}\ .$$

In particular $K\ C_0^\infty(\Omega_2) \subset C^\infty(\Omega_1)$ *if* sing spec K *contains no point which is normal to a manifold* $x = \text{constant}$.

The proof is an easy exercise and can be found in [7, p. 129]. The last statement is of course just a slight extension of a special case of Theorem 2.4.

REMARK. When $\text{supp } K \ni (x,y) \to x \in \Omega_1$ is a proper map we can obviously allow u to be any function in $C^\infty(\Omega_2)$. In particular, when $u = 1$ we obtain an estimate for the singular spectrum of the integral of K with respect to the variables in Ω_2.

If $\text{sing spec}_{\Omega_1} K = \emptyset$ we know by Theorem 2.6 and the closed graph theorem that K is a continuous map from $C_0^\infty(\Omega_2)$ to $C^\infty(\Omega_1)$. The adjoint ${}^tK$ is therefore a continuous map from $\mathcal{E}'(\Omega_1)$ to $\mathcal{D}'(\Omega_2)$. This proves the first part of the following

THEOREM 2.7. *If* $\text{sing spec}_{\Omega_2} K = \emptyset$, *then the map* K *can be extended to a continuous map from* $\mathcal{E}'(\Omega_2)$ *to* $\mathcal{D}'(\Omega_1)$. *If also* $\text{sing spec}_{\Omega_1} K = \emptyset$ *then*

$$\text{sing spec } Ku \subset \text{sing spec}' K \circ \text{sing spec } u, \; u \in \mathcal{E}'(\Omega_2)$$

where

$$\text{sing spec}' K = \{(x, \xi, y, -\eta); (x, \xi, y, \eta) \in \text{sing spec } K\}$$

is regarded as a relation mapping sets in $T^*(\Omega_2) \setminus 0$ *to sets in* $T^*(\Omega_1) \setminus 0$.

For the proof see [7, p. 130-131].

The notion of singular spectrum allows one to localize various properties of distributions $u \in \mathcal{D}'(\Omega)$ in $T^*(\Omega) \setminus 0$ or rather in the cosphere bundle $S^*(\Omega)$ consisting of rays in $T^*(\Omega) \setminus 0$. For example, just as a distribution u is in L^q at a point x_0 if $u = u_0 + u_1$ with $u_1 \in L^q$ and $x_0 \notin \text{supp } u_0$, we can say that $u \in L^q$ at (x_0, ξ_0) if $u = u_0 + u_1$ with $u_1 \in L^q$ and $(x_0, \xi_0) \notin \text{sing spec } u_0$. A case of greater interest here is obtained from the spaces $H_{(s)}(\mathbf{R}^n)$ of tempered distributions u in $\mathbf{R}^n$ with $u \in L^2_{\text{loc}}$ and

$$\int |\hat{u}(\xi)|^2 (1 + |\xi|^2)^s \, d\xi < \infty .$$

If s is a positive integer this means that the derivatives of u of order $\leqq s$ are all in L^2. It is easily seen that $C_0^\infty H_{(s)} \subset H_{(s)}$, so if $u \in H_{(s)}$ at (x_0, ξ_0) it follows that

$$\int_V |\widehat{\phi u}(\xi)|^2 (1 + |\xi|^2)^s \, d\xi < \infty ,$$

for some conic neighborhood V of ξ_0 if $\text{supp } \phi$ is close to x_0. Conversely, if $\phi(x_0) \neq 0$ this implies that $u \in H_{(s)}$ at (x_0, ξ_0).

As has been done by Sato for hyperfunctions one could formalize the preceding localization procedure by introducing the presheaf $\mathcal{C}$ of "microdistributions" on $S^*(\Omega)$ such that if O is open in $S^*(\Omega)$ and Γ is the cone generated by $\complement O$ then

$$\mathcal{C}_O = \mathcal{D}'(\Omega)/\mathcal{D}'_\Gamma(\Omega) .$$

However, we shall not use this concept explicitly here.

In the preceding discussion one can replace C^∞ by smaller classes of functions defined by conditions of the Denjoy-Carleman type. (see e.g. [9].) The case of real analytic functions is natural in the context of hyperfunctions and will occur in the lectures by M. Kashiwara.

3. *The non-characteristic regularity theorem*

Let

$$P = P(x, D) = \sum_{|\alpha| \leq m} a_\alpha(x) D^\alpha$$

be a linear partial differential operator of order m with coefficients $a_\alpha \in C^\infty(\Omega)$ where Ω is an open set in $\mathbf{R}^n$. Here $\alpha = (\alpha_1, \cdots, \alpha_n)$ is an n-tuple of non-negative integers, called a multi-index, and

$$D^\alpha = \prod_1^n (-i\,\partial/\partial x_j)^{\alpha_j} .$$

It follows immediately from the definition of sing spec u that sing spec $D^\alpha u \subset$ sing spec u, hence that sing spec $a_\alpha D^\alpha u \subset$ sing spec u and

(3.1) $$\text{sing spec } P\,u \subset \text{sing spec } u,\ u \in \mathcal{D}'(\Omega) .$$

Let $P_m(x,\xi) = \sum_{|\alpha| = m} a_\alpha(x)\xi^\alpha$ be the principal part (symbol) of P and set

$$\text{Char } P = \{(x,\xi) \in \Omega \times (\mathbf{R}^n \setminus 0);\ P_m(x,\xi) = 0\} .$$

Then we have a result in the direction opposite to (3.1).

THEOREM 3.1. *For every* $u \in \mathcal{D}'(\Omega)$ *we have*

(3.2) $$\text{sing spec } u \subset \text{Char } P \cup \text{sing spec } Pu,\ u \in \mathcal{D}'(\Omega) .$$

COROLLARY 3.2. *If* P *is elliptic, that is,* $P_m(x,\xi) \neq 0$ *in* $\Omega \times (\mathbf{R}^n \setminus 0)$ *then*

$$\text{sing spec } u = \text{sing spec } Pu, \ u \in \mathcal{D}'(\Omega) .$$

Hence sing supp u = sing supp Pu *then.*

Other applications follow by combination of Theorem 3.1 with the results of section 2. If $Pu \in C^\infty$ we conclude for example using Theorem 2.4 that the restriction of u to any submanifold with non-characteristic normal bundle is well defined. (This is a well-known result on so-called partial ellipticity.)

Proof of Theorem 3.1. If $P_m(x_0, \xi_0) \neq 0$ we can choose $\phi \in C_0^\infty(\Omega)$ with $\phi(x_0) = 1$ and an open cone $V \ni \xi_0$ so that for some $C > 0$

$$|P_m(x,\xi)| \geqq C|\xi|^m \ \text{ if } \ (x,\xi) \in \text{supp } \phi \times V .$$

We shall estimate $\widehat{\phi u}(\xi)$ when $\xi \in V$. To do so we first observe that if

$$^tP = \sum (-D)^\alpha a_\alpha$$

is the formal adjoint of P, then $Pu = f$ means that

$$\langle u, {}^tPv\rangle = \langle f, v\rangle , \ v \in C_0^\infty(\Omega) .$$

We would like to find v with ${}^tPv(x) = \phi(x)e^{-i\langle x,\xi\rangle}$. Writing

$$v(x) = e^{-i\langle x,\xi\rangle} w(x)/P_m(x,\xi)$$

we have to solve an equation of the form

$$w - Rw = \phi$$

where $R = R_1 + \cdots + R_m$ and $R_j|\xi|^j$ is a differential operator of order $\leqq j$ with C^∞ coefficients which are homogeneous of degree 0 with respect to $\xi \in V$. Formally a solution would be given by the Neumann series

$\sum_0^\infty R^k\phi$. However, the sum may not converge so we break it off and introduce

$$w_N = \sum_{k<N} R^k\phi .$$

Then we have

$$w_N - Rw_N = \phi - R^N\phi ,$$

and R^N is a sum of terms each containing $|\xi|^{-k}$ as a factor for some $k \geqq N$. The preceding equation means that

$${}^tP(x, D)(e^{-i<x,\xi>} w_N(x)/P_m(x,\xi)) = e^{-i<x,\xi>}(\phi - R^N\phi) .$$

Hence

$$\widehat{\phi u}(\xi) = u(e^{-i<.,\xi>} R^N\phi) + f(e^{-i<.,\xi>} w_N/P_m(.,\xi)),\ \xi \in V . \tag{3.3}$$

If μ is the order of the distribution u in a neighborhood of $\operatorname{supp}\phi$ then the first term on the right hand side of (3.3) can be estimated by

$$C \sum_{|\alpha| \leqq \mu} \sup |D^\alpha(e^{-i<.,\xi>} R^N\phi)| \leqq C_1 |\xi|^{\mu-N}$$

so it decreases as fast as we please if N is chosen large. If $(x_0, \xi_0) \notin$ sing spec f, the proof of Lemma 2.1 shows that one can choose a neighborhood U of x_0 and a conic neighborhood W of ξ_0 so small that for $k = 1, 2, \cdots$ and an integer M independent of k

$$\sup_W |\xi|^k |\widehat{\psi f}(\xi)| \leqq C_k \sum_{|\alpha| \leqq k+M} \sup |D^\alpha\psi|,\ \psi \in C_0^\infty(U) .$$

In fact, $(1+|\eta|)^{k+M} |\hat{\psi}(\eta)|$ can be estimated by the sum on the right. With $\phi \in C_0^\infty(U)$ and w_N defined as above we set $\psi = w_N/P_m(.,\xi)$ and conclude that the second term on the right hand side of (3.3) is $\mathcal{O}(|\xi|^{-k})$ as

$\xi \to \infty$ in W. Hence

$$\widehat{\phi u}(\xi) = \mathcal{O}(|\xi|^{-k}),\ \xi \in V \cap W,\ k = 1, 2, \cdots$$

which proves Theorem 3.1.

There is a gap between (3.1) and (3.2) when P is not elliptic. For example, if $Pu = f \in C^\infty$ we only know from Theorem 3.1 that sing spec u is a subset of the characteristics of P, and this information may be very incomplete in several respects. For example, if P has constant coefficients it is well known that one can replace Char P in (3.2) by $\Omega \times Z$ where Z is the set of all limits of $t\zeta/|\zeta|$ where $t > 0$ and $\zeta \to \infty$ on the surface $P(\zeta) = 0$ while $\operatorname{Im} \zeta$ remains bounded; if P is the heat equation this is the empty set. Even if one can find distributions u with $Pu \in C^\infty$ and sing spec u = Char P, it is not always possible to find one for which sing spec u is an arbitrary subset. The following is an elementary but instructive example:

EXAMPLE 3.2. If $D_n u = 0$ then u can be regarded (locally) as the pullback of a distribution v in $\mathbf{R}^{n-1}$ with the map $\mathbf{R}^n \ni (x', x_n) \to x' \in \mathbf{R}^{n-1}$, so

$$\text{sing spec } u \subset \{(x', x_n, \xi', 0);\ (x', \xi') \in \text{sing spec } v\}\ .$$

Here equality must hold since v is the pullback of u with the map $x' \to (x', x_n)$ for any fixed x_n. It follows that sing spec u can be any closed conic subset of the characteristic set $\xi_n = 0$ which is invariant under translations in the x_n direction, and no other sets can occur.

Our aim is to establish a general form of this result. However, before doing so it is useful to take another look at the proof of Theorem 3.1. The main point there was that the equation ${}^tP(x, D)v = \phi e^{i\langle x, \xi\rangle}$ for a fixed $\phi \in C_0^\infty$ has an asymptotic solution of the form

$$v(x) = e^{i\langle x, \xi\rangle}\, a(x, \xi)$$

where $a(x, \xi)$ is a partial sum of high order of a series

$$\sum_{j=0}^{\infty} a_{-m-j}(x,\xi)$$

where a_{-m-j} is homogeneous in ξ of degree $-m-j$. By Fourier's inversion formula we have for $f \in C_0^\infty$

$$f(x) = (2\pi)^{-n} \int e^{i<x,\xi>} \hat{f}(\xi)\, d\xi$$

so we expect to obtain an approximate solution of the equation ${}^tP(x,D)\,v = \phi f$ by taking

$$v(x) = (2\pi)^{-n} \int a(x,\xi)\, e^{i<x,\xi>} \hat{f}(\xi)\, d\xi\ .$$

This leads to the notion of pseudo-differential operator.

4. *Pseudo-differential operators*

Let Ω be an open set in $\mathbf{R}^n$ and let a be a C^∞ function in $\Omega \times \mathbf{R}^n$ which has an asymptotic expansion in homogeneous terms,

$$a(x,\xi) \sim \sum_0^\infty a_{m-j}(x,\xi) \tag{4.1}$$

where a_{m-j} is homogeneous in ξ of degree $m-j$ and C^∞ in $\Omega \times (\mathbf{R}^n \setminus 0)$. The precise meaning of (4.1) is that for every compact set $K \subset \Omega$ and arbitrary J, α, β

$$|D_\xi^\alpha D_x^\beta (a(x,\xi) - \sum_{j<J} a_{m-j}(x,\xi))| \leqq C_{K,J,\alpha,\beta}(1+|\xi|)^{m-J-|\alpha|},\ x \in K,\ |\xi| > 1\ . \tag{4.2}$$

Then a pseudo-differential operator $a(x,D)$ of order m is defined by

$$a(x,D)\,u(x) = (2\pi)^{-n} \int a(x,\xi)\, e^{i<x,\xi>} \hat{u}(\xi)\, d\xi,\ u \in C_0^\infty(\Omega)\ .$$

Since $\hat{u}$ is rapidly decreasing it is clear that $a(x, D)$ is a continuous map from $C_0^\infty(\Omega)$ to $C^\infty(\Omega)$. We have

$$e^{-i\langle x,\xi\rangle} a(x, D)(u(x) e^{i\langle x,\xi\rangle}) = (2\pi)^{-n} \int a(x, \xi+\theta) e^{i\langle x,\theta\rangle} \hat{u}(\theta)\, d\theta .$$

The contribution to the integral when $|\theta| > |\xi|/2$ is smaller than any power of $1/|\xi|$ as $\xi \to \infty$ since $a(x, \xi+\theta)$ can be estimated by $C(1+|\theta|)^{|m|}$ then and $\hat{u}$ is rapidly decreasing. Hence Taylor's formula gives

$$e^{-i\langle x,\xi\rangle} a(x, D)(u(x) e^{i\langle x,\xi\rangle}) = a_u(x, \xi) \tag{4.3}$$

where $a_u \in C^\infty$ and

$$a_u(x, \xi) \sim \sum (iD_\xi)^\alpha a(x, \xi) D_x^\alpha u(x)/\alpha! ; \tag{4.4}$$

in the right hand side we may introduce the expansion (4.1) of a. If $v \in C_0^\infty$ it follows that

$$\langle a(x, D)(u e^{i\langle \cdot,\eta\rangle}), v e^{i\langle \cdot,\xi\rangle}\rangle = \int e^{i\langle x,\xi+\eta\rangle} a_u(x, \eta) v(x)\, dx$$

decreases faster than any power of $|\xi| + |\eta|$ in a cone where $|\xi| + |\eta| < C|\xi+\eta|$ for some C. On the other hand, if $|\xi+\eta| < (|\xi| + |\eta|)/2$ we have $|\xi|/3 < |\eta| < 3|\xi|$ so the same conclusion is true if v and u have disjoint supports since a_u will be smaller than any power of $1/|\eta|$ in $\operatorname{supp} v$ then. These observations show that for the distribution kernel $A \in \mathcal{D}'(\Omega\times\Omega)$ of $a(x, D)$ we have

$$\text{sing spec } A \subset \{(x, \xi, x, -\xi);\ x \in \Omega,\ \xi \in \mathbf{R}^n \setminus 0\} ,$$

so it follows from Theorem 2.7 that $a(x, D)$ maps $\mathcal{E}'$ to $\mathcal{D}'$ and that (3.1) can be generalized to

$$\text{sing spec } a(x, D)u \subset \text{sing spec } u . \tag{4.5}$$

The proof even shows that

(4.6) $$\text{sing spec } A = \{(x,\xi;\, x,-\xi);\ x \in \Omega,\ \xi \in \mathbf{R}^n \setminus 0,\ a \text{ is not rapidly decreasing in a conic neighborhood of } (x,\xi)\}\,.$$

Given any formal sum as in (4.1) with $a_{m-j} \in C^\infty(\Omega \times (\mathbf{R}^n \setminus 0))$ homogeneous of degree $m-j$ in the second variable, it is possible to find a function $a \in C^\infty$ such that (4.1) is valid. One just has to choose $\chi \in C^\infty$ equal to 0 in a neighborhood of 0 and 1 in a neighborhood of ∞ and set

$$a(x,\xi) = \sum \chi(\xi/R_j)\, a_{m-j}(x,\xi)$$

where $R_j \to \infty$ fast enough to guarantee convergence and (4.2). Any other a satisfying (4.1) must differ from this sum by a function e which is rapidly decreasing with all its derivatives as $\xi \to \infty$, so the corresponding operator $e(x,D)$ has a C^∞ kernel

$$(2\pi)^{-n} \int e(x,\xi)\, e^{i\langle x-y,\xi\rangle}\, d\xi$$

and $e(x,D)$ maps $\mathcal{E}'$ to C^∞. Such operators are negligible when one studies singularities.

It is always possible to choose a so that both projections from the support $\subset \Omega \times \Omega$ of the distribution kernel A of a to Ω are proper, which means that $a(x,D)$ maps $\mathcal{E}'$ to $\mathcal{E}'$ and can be extended to a map from $\mathcal{D}'$ to $\mathcal{D}'$. One calls $a(x,D)$ properly supported then. To do so we just choose a partition of unity $1 = \sum \phi_j$ in Ω and $\psi_j \in C_0^\infty$ with $\psi_j = 1$ in a neighborhood of $\operatorname{supp} \phi_j$ so that no compact set meets more than finitely many $\operatorname{supp} \psi_j$. We have seen that

$$u \to \psi_j\, a(x,D)\,(\phi_j u)$$

is a pseudodifferential operator with symbol $b_j(x,\xi)$ vanishing outside $\operatorname{supp} \psi_j$, and the map $u \to b_j u - a(\phi_j u)$ has a rapidly decreasing symbol and C^∞ kernel. Hence $a' = \sum b_j$ has the required properties.

It follows easily from (4.3), (4.4) that the composition of two properly supported pseudo-differential operators $b(x, D)$ and $a(x, D)$ is another such operator $c(x, D)$ and that

$$c(x,\xi) \sim \sum (iD_\xi)^\alpha \, b(x,\xi)\, D_x^\alpha \, a(x,\xi)/\alpha\,!$$

where again we can introduce the asymptotic series for a and for b. (For a detailed proof along these lines see [6, p. 143-148].) If $a \sim \sum a_{m-j}$ and $b \sim \sum b_{\mu-j}$ then the highest order term in c is $a_m b_\mu$. Given a and a non-characteristic point (x_0, ξ_0), $\xi_0 \neq 0$, that is, $a_m(x_0, \xi_0) \neq 0$, it is always possible to find b of order $-m$ so that the symbol of the composition minus 1 is rapidly decreasing in a conic neighborhood V of (x_0, ξ_0). In fact, choose V so that $a_m(x,\xi) \neq 0$ in the closure. If we write $b \sim \sum_0^\infty b_{-m-j}$ the symbol of $b(x, D)\, a(x, D)$ starts with $b_{-m} a_m$ so we first choose b_{-m} so that

$$b_{-m} = a_m^{-1} \quad \text{in } V$$

which can be done by multiplication of a_m^{-1} by a suitable cutoff function on $\Omega \times S^{n-1}$. The term in $b(x, D)\, a(x, D)$ of order $-j$ is a sum of $b_{-m-j}\, a_m$ and a homogeneous function q of order $-j$ determined by $b_{-m}, \cdots, -b_{-m-j+1}$ which we assume already chosen. If we take $b_{-m-j} = -q/a_m$ in V the inductive choice continues, and the statement is proved.

Theorem 3.1 can now be extended to properly supported pseudo-differential operators:

$$\text{sing spec } u \subset \text{sing spec } au \cup \text{Char } a\,. \tag{4.7}$$

In fact, if $(x_0, \xi_0) \notin \text{Char } a$ we can choose a properly supported pseudo-differential operator $b(x, D)$ such that $b(x, D)\, a(x, D) = \text{identity} + c(x, D)$ where c is of order $-\infty$ in a conic neighborhood of (x_0, ξ_0). Then

$$u = b(x, D)\, a(x, D)\, u - c(x, D)\, u\,,$$

sing spec $b(x, D)\, a(x, D)\, u \subset$ sing spec $a(x, D)\, u$ by (4.5), and (4.6) implies that $(x_0, \xi_0) \notin$ sing spec $c(x, D)\, u$.

From (4.7) it follows that

$$\operatorname{sing\,spec} u \subset \bigcap_{a(x,D)u \in C^\infty} \operatorname{Char} a\,.$$

There is equality here, for assume that $(x_0, \xi_0) \notin \operatorname{sing\,spec} u$. Then we can choose $a(x,\xi)$ with $a_m(x_0, \xi_0) \neq 0$ so that a is identically 0 outside a conical neighborhood of (x_0, ξ_0) which does not meet $\operatorname{sing\,spec} u$, and it follows from (4.6) that $au \in C^\infty$. Thus

$$\operatorname{sing\,spec} u = \bigcap_{au \in C^\infty} \operatorname{Char} a \tag{4.8}$$

is an alternative definition of $\operatorname{sing\,spec} u$.

To conclude this presentation of pseudo-differential operators we list a few additional important facts:

1) The adjoint ${}^t a$ (or the adjoint a^* with respect to a sesquilinear scalar product) of $a(x, D)$ is also a pseudo-differential operator if $a(x, D)$ is a properly supported pseudo-differential operator. The symbol satisfies

$${}^t a(x,\xi) \sim \sum (iD_x D_\xi)^\alpha\, a(x, -\xi)/\alpha!\,,$$

$$a^*(x,\xi) \sim \sum (iD_x D_\xi)^\alpha\, \overline{a(x,\xi)}/\alpha!\,.$$

2) Taking into account how the error term in (4.4) depends on u it is easy to extend (4.3) to non-linear functions instead of $\langle x, \xi\rangle$. If ϕ is real valued and $d\phi \neq 0$ in $\operatorname{supp} u$ then

$$e^{-it\phi(x)}\, a(x, D)\,(u(x)\, e^{it\phi(x)}) = a_{\phi,u}(x, t) \tag{4.3$'$}$$

where

$$a_{\phi,u}(x,t) \sim \sum (iD_\xi)^\alpha\, a(x,\xi)_{\xi = t\phi'(x)}\, D_y^\alpha (u(y)\, e^{it\phi(y,x)})_{/y=x}/\alpha!$$

where $\phi(y, x) = \phi(y) - \phi(x) - \langle y-x, \phi'(x)\rangle$ is the non-linear part of ϕ at x. From this formula one deduces easily that pseudo-differential operators are invariant under a change of variables. Hence pseudo-differential operators can be defined on a C^∞ manifold, and the principal symbol transforms as a function invariantly defined on the cotangent bundle.

3) Pseudo-differential operators of order 0 map $L^2 \cap \mathcal{E}'(\Omega)$ into $L^2_{loc}(\Omega)$. To prove this we may assume $a(x,D)$ properly supported and choose a function $A \in C^\infty$ with $A(x) > a(x,\xi)$ for all ξ. Using the formulas for multiplication it is easy to construct a pseudo-differential operator $b(x,D)$ of order 0 such that

$$A(x)^2 = b(x,D)^* b(x,D) + a(x,D)^* a(x,D) + c(x,D)$$

where c is of order $-\infty$. Since

$$\|Au\|^2 \geqq \|a(x,D)u\|^2 + (c(x,D)u,u)$$

the assertion follows. Now we defined $H_{(s)}(\mathbf{R}^n)$ as the set of all u with $(1+|D|^2)^{s/2}u \in L^2$. From this it follows easily that a distribution u is in $H_{(s)}$ at (x_0,ξ_0) if and only if $a(x,D)u \in L^2_{loc}$ for every pseudo-differential operator a of order s with symbol of order $-\infty$ outside a sufficiently small conic neighborhood of (x_0,ξ_0). If $u \in H_{(s)}$ at (x_0,ξ_0) and a is of order m it follows that $a(x,D)u \in H_{(s-m)}$ at (x_0,ξ_0); conversely $a(x,D)u \in H_{(s-m)}$ at (x_0,ξ_0) implies $u \in H_{(s)}$ at (x_0,ξ_0) if this is not a characteristic point for a.

4) With rather small modifications the results concerning pseudo-differential operators discussed here remain valid for more general symbols, for example symbols in the class $S^m_{\rho,\delta}$ defined by

$$|D_\xi^\alpha D_x^\beta a(x,\xi)| \leqq C_{K,\alpha,\beta}(1+|\xi|)^{m-\rho|\alpha|+\delta|\beta|}, \; x \in K \subset\subset \Omega ,$$

where $0 \leqq 1-\rho \leqq \delta < \rho \leqq 1$.

Proofs of all results mentioned in this section can be found in [6] or [7], for example.

5. *Bicharacteristics and symplectic geometry*

Let us now return to the problem encountered in section 3 of determining which subsets of Char P that can occur as sing spec u when $Pu \in C^\infty$. We allow P to be a pseudo-differential operator of order m now but assume that the principal symbol p_m is real. Pseudo-differential operators allow us often to decrease sing spec u, for if Q is a pseudo-differential operator such that the commutator

$$[P, Q] = PQ - QP$$

is of order $-\infty$ then $Pu \in C^\infty$ implies $Pv \in C^\infty$ if $v = Qu$. We have sing spec u $\subset$ sing spec v and the inclusion is strict if the symbol of Q is rapidly decreasing in a conic neighborhood of a point in sing spec u.

If the symbol of P is $p_m + p_{m-1} + \cdots$ and that of Q is $q_\mu + q_{\mu-1} + \cdots$ then the principal symbol of $[P, Q]$ is $-i\{p_m, q\}$ where

$$\{p, q\} = \sum (\partial p/\partial\xi_j\, \partial q/\partial x_j - \partial p/\partial x_j\, \partial q/\partial \xi_j)$$

is the *Poisson bracket* of p and q. We can regard $\{p, q\}$ as a first order differential operator $H_p q = \langle H_p, dq \rangle$ acting on q, where H_p is the Hamilton vector field

$$H_p = \sum (\partial p/\partial\xi_j\, \partial/\partial x_j - \partial p/\partial x_j\, \partial/\partial\xi_j) .$$

Since $H_p p = \{p, p\} = 0$ it is clear that p is constant on integral curves of p; those on which $p = 0$ are called *bicharacteristics*. (Warning: this term is sometimes applied to all integral curves.) The condition for $[P, Q]$ to vanish of order $-\infty$ is now a sequence of equations

$$\{p_m, q_\mu\} = 0, \cdots, \{p_m, q_{\mu-j}\} = f_j$$

where f_j is homogeneous of degree $m+\mu-j-1$ and determined by $q_\mu, \cdots, q_{\mu-j+1}$. Assume for example that there is a bicharacteristic curve γ with projection in Ω approaching $\partial\Omega$ in both directions. Then it cannot be tangent to the radial vector field $\langle \xi, \partial/\partial\xi \rangle$ at any point, for it would just move in that direction forever then. Hence we can prescribe homogeneous initial data on a conic hypersurface transversal to the curve γ at one point and obtain for a relatively compact open set $\Omega' \subset\subset \Omega$ solutions of all these equations vanishing outside an arbitrarily small conic neighborhood V of γ while $q_\mu \neq 0$ on γ. But then $Qu \subset C^\infty(\Omega')$ and sing spec $Qu \subset V$ while $\gamma \cap$ sing spec $u = \gamma \cap$ sing spec Qu. It follows that if there is an extension of Example 3.2 valid for P, then it must say that the singular spectrum of a distribution u with $Pu \in C^\infty$ is invariant under the Hamilton flow defined by the vector field H_p in the characteristic set. This is indeed true:

THEOREM 5.1. *Assume that* P *is a properly supported pseudo-differential operator with real principal symbol* p. *If* $u \in \mathcal{D}'$ *and* $Pu = f$ *it follows then that* sing spec u \ sing spec f *is contained in* $p^{-1}(0)$ *and invariant under the flow defined by the Hamilton vector field* H_p.

The theorem could be proved by combining the preceding arguments with classical results on hyperbolic operators (see [8]). However, instead of doing so we shall develop a general transformation theory which allows us to reduce the proof of the theorem to Example 3.2.

The Poisson bracket $\{p, q\}$ above can also be regarded as a bilinear form $\sigma(H_p, H_q)$ where σ is the symplectic form

$$\sigma = \sum d\xi_j \wedge dx_j = d\left(\sum \xi_j dx_j\right).$$

The one form $\sum \xi_j dx_j$ is invariantly defined on the cotangent bundle of any manifold Ω; it is characterized by the fact that its pullback to Ω under the map $\Omega \ni x \to (x, \phi'(x))$ is just $d\phi$ if ϕ is a function on Ω. Replacing H_p by t above which can then be any tangent vector to $T^*\Omega$ we have

$$\langle t, dq \rangle = \sigma(t, H_q)$$

which shows that the definition of H_q only depends on the symplectic form. Hence it makes sense if Ω is a manifold which was anticipated in stating Theorem 5.1 for a manifold. More generally, if Ω' is another manifold of the same dimension as Ω and $\chi: T^*(\Omega') \to T^*(\Omega)$ is a map such that $\chi^*\sigma_\Omega = \sigma_{\Omega'}$ then the bicharacteristics of χ^*p in $T^*(\Omega')$ are mapped by χ on those of p in $T^*(\Omega)$. Such a map χ is called *canonical*. (Note that since σ_Ω^n is a volume form in Ω, if n is the dimension of Ω, and $\chi^*\sigma_\Omega^n = \sigma_{\Omega'}^n$, a canonical map is automatically a local diffeomorphism.) We must of course require that χ^*p is homogeneous of the same order as p so we also want χ to commute with the multiplication by positive scalars in $T^*(\Omega)$ and $T^*(\Omega')$.

Finding a canonical transformation $T^*(\Omega) \to T^*(\mathbf{R}^n)$ (locally) with the standard coordinates in $T^*(\mathbf{R}^n)$ means to determine functions $x_j = q_j$ and $\xi_j = p_j$ on $T^*(\Omega)$ which are homogeneous of degree 0 and 1 respectively and satisfy the commutation relations

$$\{p_j, p_k\} = \{q_j, q_k\} = 0, \ \{p_j, q_k\} = \delta_{jk} \ \text{ for } j, k = 1, \cdots, n \,. \tag{5.1}$$

The problem of reducing the study of solutions of the equation $Pu = 0$ to the study of solutions of the equation $D_n u = 0$ in Example 3.2 can roughly speaking be split into two: a) The geometrical problem of finding a canonical transformation which reduces the principal symbol to ξ_n, and b) The analytical problem of finding a corresponding map F such that $FPF^{-1} = D_n$. (We must assume then that P is of first order which is no serious restriction since we can always multiply P by an elliptic operator of order $1-m$ without changing any singularities.) The first problem is easily solved by the homogeneous Darboux theorem which we now state in a form which is sufficiently general for other applications also:

THEOREM 5.2. *Let Ω be an n dimensional manifold, A and B two subsets of $1, \cdots, n$ and q_α, $\alpha \in A$; p_β, $\beta \in B$, C^∞ functions in a conic neighborhood of a point $\gamma \in T^*(\Omega)\setminus 0$ such that*

(i) *q_α and p_β are homogeneous of degree 0 and 1 respectively.*

(ii) $\{q_\alpha, q_{\alpha'}\} = \{p_\beta, p_{\beta'}\} = 0$; $\{p_\beta, q_\alpha\} = \delta_{\alpha\beta}$ *if* $\alpha, \alpha' \in A$ *and* $\beta, \beta' \in B$.

(iii) *The differentials* dq_α, dp_β *and the canonical one form* $\omega = \langle \xi, dx \rangle$ *are linearly independent at* γ.

Let a_α, b_β *be arbitrary real numbers,* $\alpha \notin A$, $\beta \notin B$, *such that there is some* $\alpha_0 \notin A$ *for which* $p_{\alpha_0}(\gamma) \neq 0$ *in case* $\alpha_0 \in B$ *and* $b_{\alpha_0} \neq 0$ *otherwise. Then one can find* q_α, $\alpha \notin A$ *and* p_β, $\beta \notin B$ *such that* (i) *and* (ii) *remain valid for all indices,*

$$p_\beta(\gamma) = b_\beta,\ \beta \in B;\ q_\alpha(\gamma) = a_\alpha,\ \alpha \in A\ ,$$

and (iii) *remains valid when* α *and* β *are arbitrary indices with* $\alpha \neq \alpha_0$.

An only slightly weaker result was given in Melrose [10] (see also [2, Theorem 6.1.3]). One chooses all p_β first, then all q_α with q_{α_0} the last to be chosen. The homogeneity conditions can be written as differential equations

$$\lambda q_\alpha = 0,\ \lambda p_\beta = p_\beta$$

where $\lambda = \langle \xi, \partial/\partial\xi \rangle$ is the radial vector field. Each choice then means the solution of a system of differential equations in involution and one just has to take care to preserve the condition (iii) which makes the leading terms of the differential equations linearly independent.

In particular, if we are just given a real valued function p which is homogeneous of degree one and for which H_p is not in the radial direction at γ then we can take $p_n = p$ and obtain a canonical transformation which transforms p to the coordinate ξ_n in $T^*(\mathbf{R}^n)$. This solves the geometrical problem a) stated above in this case. The analytical problem b) will be postponed until the next section but we shall now discuss the definition of the canonical transformation by means of a *generating function.*

Suppose that $(y_0, \eta_0) \in T^*(\mathbf{R}^n) \setminus 0$ and $(x_0, \xi_0) \in T^*(\Omega) \setminus 0$, where $\dim \Omega = n$, and that S is a C^∞ function defined in a neighborhood of $(x_0, \eta_0) \in \Omega \times \mathbf{R}^n$ such that $\det \partial^2 S/\partial x \partial\eta \neq 0$ at (x_0, η_0). Then

$\chi(y,\eta) = (x,\xi)$ where

$$\xi = \partial S(x,\eta)/\partial x, \quad y = \partial S(x,\eta)/\partial \eta$$

defines a canonical transformation from a neighborhood of (y_0, η_0) to a neighborhood of (x_0, ξ_0), which is homogeneous if S is homogeneous with respect to η of degree 1. In fact, by the implicit function theorem the equation $y = \partial S(x,\eta)/\partial \eta$ defines x as a function of y and η so we obtain a well-defined map χ locally. Furthermore,

$$\sum d\xi_j \wedge dx_j = \sum \partial^2 S/\partial \eta_k \partial x_j \, d\eta_k \wedge dx_j = \sum d\eta_k \wedge dy_k$$

since the matrices $\partial^2 S/\partial x_k \partial x_j$ and $\partial^2 S/\partial \eta_k \partial \eta_j$ are symmetric. S is called a generating function of χ. Conversely, if χ is a canonical transformation of a neighborhood of (y_0, η_0) on a neighborhood of (x_0, ξ_0) such that with $\chi(y,\eta) = (x,\xi)$ we have $\det \partial x/\partial y \neq 0$ at (y_0, η_0) then we can regard ξ and y as functions of x and η. The form

$$\sum \xi_j \, dx_j + \sum y_j \, d\eta_j$$

is closed since $\sum d\xi_j \wedge dx_j + \sum dy_j \wedge d\eta_j = 0$. Thus it can locally be represented as $dS(x,\eta)$ where S is then a generating function, uniquely determined up to a constant term. When the canonical transformation is homogeneous we obtain a homogeneous S by taking $S(x,\eta) = \sum \eta_j y_j$, for $\sum \eta_j \, dy_j = \sum \xi_j \, dx_j$ then. In fact, χ preserves the radial vector field $\sum \xi_j \partial/\partial \xi_j$ then and $\sum \xi_j \, dx_j$ is just the inner product of the symplectic form $\sum d\xi_j \wedge dx_j$ with it.

The condition $\det \partial x/\partial y \neq 0$ means that there is no horizonthal vector $\neq 0$ in $T^* \mathbf{R}^n$, that is, one with $d\eta = 0$, which is mapped to a tangent of the fiber of $T^*(\Omega)$ at (x_0, ξ_0). Now the notion of horizonthal vector depends entirely on the fact that coordinates are given in $\mathbf{R}^n$; the horizonthal vectors at (y_0, η_0) form the tangent plane of the section

$\{y, \phi'(y)\}$ of $T^*(\mathbf{R}^n)$ defined by the linear function with $\phi'(y) = \eta_0$. Now if

$$\psi(y) = \phi(y) + \langle A(y-y_0), y-y_0 \rangle/2$$

where A is a symmetric matrix, then

$$\{(y, \psi'(y))\} = \{(y, \eta_0 + A(y-y_0)\}$$

with tangent plane defined by $\eta = Ay$. This is an arbitrary *Lagrangean* plane which is transversal to the plane $y = 0$. (By Lagrangean plane one means a plane of the maximal dimension n on which the symplectic form vanishes identically.) Now we can choose A so that it is also transversal to the plane which is the inverse image under χ' of the tangent to the fiber of $T^*(\Omega)$ at (x_0, ξ_0) which is also Lagrangean since χ is symplectic. This is true for a generic A. (See e.g. [7, p. 136-137] for details.) If we take new coordinates in $\mathbf{R}^n$ so that ψ is a linear function of the new coordinates at y_0, which is possible since $\eta_0 \neq 0$, then χ has a generating function in the new variables. Summing up:

THEOREM 5.3. *If* χ *is a homogeneous canonical transformation from a conic neighborhood of* $(y_0, \eta_0) \in T^*(\Omega') \setminus 0$ *to a conic neighborhood of* $(x_0, \xi_0) \in T^*(\Omega) \setminus 0$, *then one can choose local coordinates in* Ω' *at* y_0 *so that* χ *has a homogeneous generating function in a neighborhood of* (y_0, η_0).

6. *Fourier integral operators corresponding to canonical transformations*

Let Ω and Ω' be n dimensional manifolds and χ a homogeneous canonical transformation from a conic neighborhood of $(y_0, \eta_0) \in T^*(\Omega') \setminus 0$ to a conic neighborhood of $(x_0, \xi_0) \in T^*(\Omega) \setminus 0$. We choose local coordinates at x_0 and y_0 so that χ and χ^{-1} both have generating functions. Let S be the homogeneous generating function for χ. With a symbol a of order m supported in a small conic neighborhood of (x_0, η_0) the definition of a Fourier integral operator is a natural modification of that of a pseudo-differential operator:

$$\text{(6.1)} \qquad Fu(x) = (2\pi)^{-n} \int e^{iS(x,\eta)} a(x,\eta) |\det S''_{x\eta}(x,\eta)|^{1/2} \hat{u}(\eta) d\eta .$$

Here u is assumed to have support in the local coordinate patch at y_0 and the Fourier transform $\hat{u}$ is defined by means of the local coordinates there. We want to find an analogue of (4.3)'. To do so we introduce the definition of $\hat{u}$ and obtain formally

$$Fu(x) = (2\pi)^{-n} \iint e^{i(S(x,\eta) - \langle y,\eta \rangle)} a(x,\eta) |\det S''_{x\eta}|^{1/2} u(y)\, dy\, d\eta$$

but the integral is no longer absolutely convergent. Replacing u by $u\, e^{it\phi}$ gives

$$(F\, u\, e^{it\phi})(x) = (2\pi)^{-n} \int e^{i(S(x,\eta) - \langle y,\eta \rangle + t\phi(y))} \cdot$$

$$\cdot\ a(x,\eta) |\det S''_{x\eta}(x,\eta)|^{1/2} u(y)\, dy\, d\eta .$$

By the method of stationary phase one expects the main contributions to be caused by stationary points of the exponent, and these are defined by

$$\text{(6.2)} \qquad y = S'_\eta(x,\eta), \quad \eta = t\phi'(y) .$$

The first equation implies in view of the homogeneity of S that

$$S(x,\eta) - \langle y,\eta \rangle + t\phi(y) = t\phi(y) ,$$

and since S is the generating function of χ we have $\chi(y, t\phi'(y)) = (x, S'_x)$, so x is the projection in Ω of $\chi(y, \phi'(y))$. To be able to apply the method of stationary phase one needs to know that the critical point is not degenerate, that is, that the determinant

$$\text{(6.3)} \qquad \det\begin{pmatrix} t\phi''_{yy} & -I \\ -I & S''_{\eta\eta} \end{pmatrix} = \det (S''_{\eta\eta}\, t\phi''_{yy} - I)$$

is not 0. Now we have with x defined by (6.2) as a function of y that

$$S''_{\eta x}\, \partial x/\partial y = I - S''_{\eta\eta}\, t\phi''_{yy}$$

so the non-vanishing of (6.3) means precisely that χ maps the section $\{(y, \phi'(y)\}$ of $T^*(\Omega')$ over supp u to a section of $T^*(\Omega)$. Since $\chi^*\left(\sum \xi_j dx_j\right) = \sum \eta_j dy_j$ the image is then $\{(x, \tilde{\phi}'(x)\}$ where $\tilde{\phi}(x) = \phi(y)$ when x and y are related by (6.2). Now the stationary phase method easily gives a strict proof (cf. [7, section 3.2]) that

$$(F\, u\, e^{it\phi})(x) = b(x,t)\, e^{it\tilde{\phi}(x)} \tag{6.4}$$

where $b(x,t)$ has an asymptotic expansion in homogeneous functions of t vanishing outside a small neighborhood of x_0 and with leading term

$$b_m(x,t) = a_m(x,\eta)\, e^{\pi i\sigma/4}\, |\det \partial y/\partial x|^{\frac{1}{2}}\, u(y) \tag{6.5}$$

where again η and y are functions of x given by (6.2). σ is the signature of the first matrix in (6.3). We can regard $a_m(x,\eta)$ as a function $F_m(y,\eta)$ of (y,η) vanishing outside a small conic neighborhood of $(y_0, \eta_0) \in T^*(\Omega')$ and call it the principal symbol of F. (Strictly speaking one should take the factor $e^{\pi i\sigma/4}$ into account and define F_m with values in the Maslov line bundle but this is irrelevant in our local discussion. The square root in (6.5) can be made to disappear by regarding F as an operator between densities of order ½; it will have no importance either in what follows.)

From (6.5) it follows as in the proof of (4.6) essentially that for the distribution kernel $\mathcal{F}$ of F we have (cf. Theorem 2.7) if W is conic, supp a $\subset$ W,

$$\text{sing spec}'\mathcal{F} \subset \{(x,\xi,y,\eta);\ (x,\xi) = \chi(y,\eta),\ (x,\eta) \in W\}\,. \tag{6.6}$$

Thus F defines a map from $\mathcal{E}'$ to $\mathcal{D}'$ such that

$$\text{sing spec } Fu \subset \chi(\text{sing spec } u \cap V) \tag{6.7}$$

where V is a small conic neighborhood of (y_0, η_0). As in the case of pseudo-differential operators we can make F properly supported by subtracting an operator with C^∞ kernel and will always assume this done.

If P is a proper pseudo-differential operator in Ω we can use (4.3)′ to show that it is possible to operate with P inside the integral sign in (6.1) and conclude that PF is another Fourier integral operator corresponding to the canonical transformation χ with the principal symbol of F multiplied by $p(x, S'_x)$. Since $(x, S'_x) = \chi(y, \eta)$ this means that the principal symbol of PF is $(p \circ \chi)$ times the principal symbol of F.

Now we can in the same way introduce a Fourier integral operator F_1 from functions in Ω to functions in Ω' corresponding to the inverse canonical transformation χ^{-1}. When F_1 is applied to (6.4) the oscillatory factor $e^{it\tilde{\phi}}$ is converted back to $e^{it\phi}$ so in particular

$$(F_1 F\, u\, e^{i\langle \cdot, \eta\rangle})(y) = c(y, \eta) e^{i\langle y, \eta\rangle}$$

where c has an asymptotic expansion in homogeneous functions which is of order $-\infty$ except in a conic neighborhood of (y_0, η_0) where it is the product $(f_1 \circ \chi) f$ of the principal symbols of F_1 and F times possibly a factor $e^{\pi i \sigma/4}$ which is a constant of modulus one which we ignore since it can be absorbed in F_1 say. But this means that the composition is a pseudo-differential operator with symbol c. At a non-characteristic point it preserves the singular spectrum so we conclude that

(6.8) $\quad (y, \eta) \in \text{sing spec } u \iff \chi(y, \eta) \in \text{sing spec } Fu \quad \text{if } F_m(y, \eta) \neq 0 .$

It is not hard to prove (see [7, section 4.2]) that the adjoint operator of F modulo an operator with C^∞ kernel is a Fourier integral operator belonging to χ^{-1}. (Note that it follows from (6.4) that our definition of this concept is independent of the choice of local coordinates.) This implies that multiplication by pseudo-differential operators to the right also preserves the class of Fourier integral operators belonging to χ, and it shows that operators of order 0 are continuous from $L^2 \cap \mathcal{E}'$ to L^2_{loc}

since F^*F is. More generally, if $u \in H_{(s)}$ at (y, η) then $Fu \in H_{(s)}$ at $\chi(y, \eta)$ if F is of order 0.

We have now developed the tools which allow us to complete the proof of Theorem 5.1. We keep the notations there and let (x_0, ξ_0) be a characteristic point with

$$(x_0, \xi_0) \notin \text{sing spec } Pu\,. \tag{6.9}$$

We may assume that H_p does not have the radial direction at (x_0, ξ_0) for there is nothing to prove then. By Theorem 5.2 we can then find a canonical transformation from a neighborhood of $(0, \eta_0) \in T^*(\mathbf{R}^n) \setminus 0$ to a neighborhood of (x_0, ξ_0) such that $p \circ \chi(y, \eta) = \eta_n$. Choose as above Fourier integral operators F and F_1 corresponding to χ and χ^{-1} with principal symbol different from 0 at $(0, \eta_0)$ and (x_0, ξ_0) respectively. We may assume that $FF_1 - I$ has symbol of order $-\infty$ in a conic neighborhood of (x_0, ξ_0) since this can be achieved by multiplication of F by a pseudo-differential operator of order 0 as in the proof of (4.7). Then the distribution $v = F_1 u$ has the property that $(x_0, \xi_0) \notin$ sing spec $Fv - u$, so $(x_0, \xi_0) \notin$ sing spec $PFv - Pu$ which gives $(x_0, \xi_0) \notin$ sing spec PFv in view of (6.9), hence

$$(0, \eta_0) \notin \text{sing spec } F_1 PFv\,. \tag{6.10}$$

Now $Q = F_1 PF$ is a pseudo-differential operator with symbol $q = p \circ \chi(y, \eta) = \eta_n$ near $(0, \eta_0)$, and sing spec $u = \chi$ sing spec v near (x_0, ξ_0). Thus we have reduced the proof of Theorem 5.1 to the same result for the operator Q with principal symbol D_n. The following simple lemma will eliminate the lower order terms in D_n and reduce the theorem entirely to that operator.

LEMMA 6.1. *If Q is a pseudo-differential operator with principal symbol η_n in a conic neighborhood of $(0, \eta_0)$ then one can find an elliptic pseudo-differential operator A of order 0 such that the symbol of*

$$AQ - D_nA$$

is of order $-\infty$ *in a conic neighborhood of* $(0, \eta_0)$.

Proof. If we write the symbols of A and Q as $\sum_0^\infty a_{-j}$ and $\sum_0^\infty q_{1-j}$ then the highest order term in $AQ - D_nA = [A, D_n] + A(Q - D_n)$ is

$$i\partial a_0/\partial x_n + a_0 q_0$$

and we can make it vanish in a neighborhood V of $(0, \eta_0)$ by solving a differential equation, that is, by taking

$$a_0(x, \xi) = \exp \int_0^{x_n} i\, q_0(x', t, \xi)\, dt .$$

The vanishing of the term of order $-j$ is an inhomogeneous differential equation of the same form for a_{-j} and can be solved in the same way. (For details see [2, p. 200-201].)

To A we can construct a pseudo-differential operator B such that $BA - I$ and $AB - I$ are of order $-\infty$, and this gives $D_n - AQB$ of order $-\infty$ in a conic neighborhood of $(0, \eta_0)$. Thus $AF_1PFB - D_n$ is of order $-\infty$ in a conic neighborhood of $(0, \eta_0)$ so a complete reduction of P to the form D_n is made by the Fourier integral operators FB and AF_1 which correspond to χ and to χ^{-1}. For this operator the proof of Theorem 5.1 is essentially the simple argument in Example 3.2 (cf. [2, p. 196]) so this completes the proof. In the same way one obtains a more precise result ([2, Th. 6.1.1′]):

THEOREM 6.2. *Assume that* P *is a properly supported pseudo-differential operator of order* m *with real principal symbol. If* $u \in \mathcal{D}'(\Omega)$ *and* γ *is an interval on a bicharacteristic strip where* $Pu \in H_{(s)}$ *then* $u \in H_{(s+m-1)}$ *on* γ *if this is true at some point on* γ.

We shall finally justify the statement made in the introduction that existence theorems can be derived from the study of singularities. First we give a consequence of Theorem 6.2.

THEOREM 6.3. *Let* P *be a properly supported pseudo-differential operator in* Ω *of order* m *with real principal symbol and assume that* K *is a compact subset such that no bicharacteristic for* P *stays over* K *forever. Then*

$$u \in \mathcal{E}'(K),\ Pu \in H_{(s)} \Longrightarrow u \in H_{(s+m-1)}\,.$$

Proof. We know from the end of section 4 that $u \in H_{(s+m)}$ at non-characteristic points. Through a characteristic point there is a bicharacteristic which contains some point outside K. Since $u \in H_{(s+m-1)}$ there it follows from Theorem 6.2 that the same is true at every point on the bicharacteristic. Hence $u \in H_{(s+m-1)}$ since this is true at every point in $T^*(\Omega)\setminus 0$.

In particular Theorem 6.3 shows that

$$N = \{u \in \mathcal{E}'(K),\ Pu = 0\} \subset C_0^\infty(K)\,, \tag{6.11}$$

and N must be finite dimensional since it is a locally compact Banach space with the L^2 norm for example. Theorem 6.3 gives in view of the closed graph theorem

$$\|u\|_{(s+m-1)} \leqq C(\|Pu\|_{(s)} + \|u\|_{(s+m-2)}),\ u \in C_0^\infty(K)\,, \tag{6.12}$$

and an elementary compactness argument then gives

$$\|u\|_{(s+m-1)} \leqq C'\|Pu\|_{(s)} \quad \text{if} \quad u \in C_0^\infty(K),\ \langle u, v_j\rangle = 0,\ j = 1, \cdots, J\,, \tag{6.13}$$

where $v_j \in C_0^\infty(K)$ form a biorthogonal basis to N. From the Hahn-Banach theorem one can now deduce that for every $f \in H_{(s)}(\Omega)$ orthogonal to N there is a distribution $u \in H_{(s+m-1)}(\Omega)$ with ${}^tPu = f$ in the interior of K.

Again details can be found in [2, p. 206]. The idea of simplifying pseudo-differential operators by conjugation with Fourier integral operators goes back to Egorov [4] (see also e.g. Nirenberg and Treves [13]).

7. *Further equivalence theorems*

Let P be a properly supported pseudo-differential operator of order m in Ω with principal symbol p no longer assumed to be real. Let $(x_0, \xi_0) \in T^*(\Omega)\setminus 0$ be a characteristic point where $H_p \neq 0$. Multiplying P by i if necessary we may assume that $H_{\mathrm{Re}\, p} \neq 0$ there. The equation $\mathrm{Re}\, p = 0$ then defines a smooth hypersurface through (x_0, ξ_0) foliated by the bicharacteristics of $\mathrm{Re}\, p$. Assume

(7.1) Im p has a zero of fixed order $k \neq 0$ near (x_0, ξ_0) on every bicharacteristic of Re p close to the one through (x_0, ξ_0).

This implies that $H_{\mathrm{Re}\, p}$ does not have the radial direction for Im p would then vanish identically on the bicharacteristic through (x_0, ξ_0).

THEOREM 7.1. *If* (7.1) *is valid one can find a homogeneous canonical transformation* χ *from a neighborhood of* $(0, \eta_0) \in T^*(\mathbf{R}^n)\setminus 0$ *to a neighborhood of* (x_0, ξ_0) *and Fourier integral operators* F, F_1 *corresponding to* χ *and* χ^{-1} *such that*

$$F_1 P F - (D_1 + i x_1{}^k D_2)$$

has symbol of order $-\infty$ *in a conic neighborhood of* $(0, \eta_0)$. *One can take* $\eta_{02} > 0$ *if* k *is even whereas* η_{02} *must have the sign of* $H_{\mathrm{Re}\, p}{}^k \mathrm{Im}\, p(x_0, \xi_0)$ *if* k *is odd.*

The last statement is obvious since multiplication of p by a non-vanishing function does not change the sign and since the expression is invariant under canonical transformations. The preceding theorem is due to Sato-Kawai-Kashiwara [14] in the analytic case and to Duistermaat-

Sjöstrand [3] in the C^∞ case with $k = 1$, Yamamoto [16] for the C^∞ case with $k > 1$. (His result is incomplete with respect to removal of lower order terms which requires an analogue of Lemma 6.1 with two different operators A.)

The differential operator $D_1 + ix_1{}^k D_2$ is a modification first considered by Mizohata [12] of the famous example of H. Lewy of a differential equation with no solutions.

PROPOSITION 7.2. *If* k *is odd and* $\eta_{01} = 0$, $\eta_{02} < 0$ *then one can find a* C^1 *solution of the equation* $(D_1 + ix_1{}^k D_2)u = 0$ *with* $\text{sing spec}\, u = \{(0, t\eta_0), t > 0\}$.

Thus the solution has a singularity which does not propagate.

Proof. Let $v(x')$, $x' = (x_2, \cdots, x_n)$ be a function of compact support with $\text{sing spec}\, v = \{(0, t\eta_0'), t > 0\}$. (See Example 2.3.) Then $\hat{v}$ is rapidly decreasing outside any conic neighborhood Γ of η_0'. Now set

$$u(x) = (2\pi)^{1-n} \int_\Gamma \exp\,(\xi_2 x_1{}^{k+1}/(k+1) + i\langle x', \xi'\rangle)\hat{v}(\xi')\,d\xi' .$$

We assume Γ chosen so that $\xi_2 \leqq -|\xi'|c$ in Γ. If we take v sufficiently differentiable it is clear that this gives a C^1 solution of $(D_1 + ix_1{}^k D_2)u = 0$, which is C^∞ for $x_1 \neq 0$ and differs from v by a C^∞ function when $x_1 = 0$. The equation implies that $\xi_1 = 0$ in sing spec u so it follows from Theorem 2.4 that $\text{sing spec}\, u \supset \{(0, t\eta_0), t > 0\}$. If $\phi \in C_0^\infty$, $\phi(x) = \phi_0(x_1)\,\phi_2(x')$ then

$$\widehat{\phi u}(\theta) = (2\pi)^{1-n} \int_\Gamma \hat{\phi}_2(\theta' - \xi')\hat{v}(\xi')\,d\xi' \int \phi_0(x_1)\exp\,(\xi_2 x_1{}^{k+1}/(k+1) - ix_1\theta_1)\,dx_1$$

is rapidly decreasing if $\theta' \to \infty$ in a direction outside Γ while $|\theta_1| < |\theta'|/2$ say. Writing u as a convolution one obtains easily that sing supp $u \subset \{0\}$. Hence

$$\text{sing spec } u \subset \{(0;\, 0, \eta'),\ \eta' \in \Gamma\}\ .$$

Decreasing Γ changes u only by a C^∞ term which does not affect sing spec u, so the preceding conclusion must be valid with Γ replaced by the ray through η_0' which proves the proposition.

Combination of 7.1 and Proposition 7.2 gives the first part of

THEOREM 7.3. *If* (7.1) *is valid,* k *is odd, and* $H_{\mathrm{Re}\,p}^{\,k} \operatorname{Im} p(x_0, \xi_0) < 0$, *then one can find a distribution* u *with* $Pu \in C^\infty$ *and* sing spec $u = \{(x_0, t\xi_0),\ t > 0\}$. *In the other cases where* (7.1) *is valid we have on the other hand that* $Pu \in H_{(s)}$ *at* (x_0, ξ_0) *implies* $u \in H_{(s+m-k/(k+1))}$ *at* (x_0, ξ_0).

The last statement is a very special case of the results on subellipticity by Egorov [5] (see also Treves [15]) and can of course also be proved by inspection of the Mizohata operator (cf. Duistermaat and Sjöstrand [3]). Subellipticity will be discussed at great length later on in the seminar.

For P to be *hypoelliptic* in the sense that

$$\text{sing spec } u \subset \text{sing spec } Pu,\ u \in \mathcal{D}'$$

it is therefore necessary that $\operatorname{Im} p$ never changes sign from $+$ to $-$ along a bicharacteristic of $\operatorname{Re} p$ at a zero of finite order. In fact, nearby there must then be such a zero of minimal multiplicity and then (7.1) is fulfilled.

As already emphasized repeatedly results on the regularity of solutions of a differential equation are closely related to existence theorems for the adjoint. We shall now discuss such results which also follow from Theorem 7.1 and an examination of the Mizohata operators.

DEFINITION 7.4. We shall say that the pseudo-differential operator P is solvable at (x_0, ξ_0) if for every $f \in \mathcal{D}'$ there is a distribution u such that $(x_0, \xi_0) \notin$ sing spec $(Pu-f)$.

Solvability at (x_0, ξ_0) for every ξ_0 is of course necessary for solvability of the equation $Pu = f$ in a neighborhood of x_0 for every f.

THEOREM 7.5. *If (7.1) is fulfilled and* k *is odd,* $H_{\mathrm{Re}\,p}^k \,\mathrm{Im}\, p > 0$, *then* P *is not solvable at* (x_0, ξ_0).

The proof is an easy adaptation of that of Duistermaat-Sjöstrand [3] for $k = 1$.

It follows from Theorem 7.5 that local solvability of P requires the following condition

(ψ) On a bicharacteristic of $\mathrm{Re}\, p$ there must not be any zero of finite order where $\mathrm{Im}\, p$ changes sign from $-$ to $+$, and similarly for ip.

If P is a differential operator it follows by considering the symmetric point $(x, -\xi)$, and noting that $H_{\mathrm{Re}\,p}^k \,\mathrm{Im}\, p$ is odd when k is odd, that local solvability requires the condition

(p) On a bicharacteristic of $\mathrm{Re}\, p$ there must not be any zero of finite order where $\mathrm{Im}\, p$ changes sign, and similarly for ip.

When the coefficients are real analytic it was proved by Nirenberg and Treves [13] that this condition implies local solvability provided that $H_{\mathrm{Re}\,p}$ never has the radial direction. Beals and Fefferman [1] have shown the same result in the C^∞ case assuming that (p) is strengthened to the condition (P) that there is no sign change even at zeros of infinite order which may then occur. To obtain semiglobal existence theorems similar to those at the end of section 6 one has to prove theorems on the propagation of singularities replacing Theorem 6.2. This will be discussed more fully later on in section 8 but we content ourselves now with a result from [2, Chap. VII].

THEOREM 7.6. *Assume that* P *is a pseudo-differential operator in* Ω *and that the principal symbol* $p = p_1 + ip_2$ *is characteristic at* (x_0, ξ_0) *and that* H_{p_1}, H_{p_2} *and the radial vector field are linearly independent there. If condition* (p) *is fulfilled in a neighborhood, that is,*

$$\{p_1, p_2\} = 0 \text{ when } p = 0 \, ,$$

then one can find a homogeneous canonical transformation χ *from a neighborhood of* $(0, \eta_0) \in T^*(R^n)\setminus 0$ *to a neighborhood of* (x_0, ξ_0) *and Fourier integral operators* F, F_1 *corresponding to* χ *and* χ^{-1} *such that*

$$F_1 P F - (D_1 + iD_2)$$

has symbol of order $-\infty$ *in a conic neighborhood of* $(0, \eta_0)$.

The hypothesis $\{p_1, p_2\} = 0$ when $p_1 = p_2 = 0$ implies that

$$\{p_1, p_2\} = a_1 p_1 + a_2 p_2$$

for some smooth functions a_1 and a_2. Now we use the Jacobi identity

$$\{p_1,\{p_2,p_3\}\} + \{p_3,\{p_1,p_2\}\} + \{p_2,\{p_3,p_1\}\} = 0$$

or equivalently

$$H_{\{p_1,p_2\}} = [H_{p_1}, H_{p_2}] \, .$$

This gives when $p = 0$

$$[H_{p_1}, H_{p_2}] = a_1 H_{p_1} + a_2 H_{p_2} \, .$$

By the Frobenius theorem the manifold of codimension 2 where $p = 0$ is therefore foliated by two dimensional manifolds with tangent plane spanned by H_{p_1} and H_{p_2}. It is natural to call them bicharacteristics. They have a complex structure given by regarding the complex vector field H_p as the Cauchy-Riemann operator; the bicharacteristic and the structure do not change if P is multiplied by a non-characteristic operator. The reduction given by Theorem 7.6 can now be used to show that in sing spec $u \setminus$ sing spec Pu the "regularity function"

$$s_u^*(x,\xi) = \sup\{s;\ u \in H_{(s)} \text{ at } (x,\xi)\}$$

is superharmonic along the bicharacteristics ([2, Theorem 7.2.1]). We shall see in section 8 that the result can be extended to the general situation studied by Nirenberg and Trèves [13] and Beals-Fefferman [1].

8. *Propagation of singularities and semi-global existence theorems for pseudo-differential operators satisfying condition* (P)

Let Ω be a C^∞ manifold and P a pseudo-differential operator of order m in Ω, with principal symbol denoted by p. In section 7 we saw that at least for differential operators the following condition is not far from being necessary for local solvability of P:

DEFINITION 8.1. p is said to satisfy condition (P) if there is no C^∞ complex valued function q in $T^*(\Omega)\setminus 0$ such that Im qp takes both positive and negative values on a bicharacteristic of Re qp where $q \neq 0$.

By a bicharacteristic of Re qp we mean an integral curve of the Hamilton field Re H_{qp} on which Re qp vanishes. (Some authors call this a null bicharacteristic.) A curve will be called a semibicharacteristic of p if it is a bicharacteristic of Re qp for some $q \neq 0$.

We want to prove the following extension of Theorem 6.3:

THEOREM 8.2. *Assume that* p *satisfies condition* (P) *and let* $K \subset \Omega$ *be a compact set such that every characteristic point over* K *lies on a semi-bicharacteristic whose end points are non-characteristic or located over the complement of* K. *Then if* $s \in \mathbf{R}$ *or* $s = +\infty$ *we have*

(8.1) $u \in \mathcal{E}'(K)$, $Pu \in H_{(t)}$ *for every* $t < s \Longrightarrow u \in H_{(t+m-1)}$ *for every* $t < s$.

From (8.1) we can obtain existence theorems exactly as in section 6 after Theorem 6.3. In particular, the equation ${}^tPu = f$ can be solved in a neighborhood of K for all distributions f which are orthogonal to a finite dimensional subspace of $C_0^\infty(K)$; one can find u with m–1 derivatives more than f. This extends the results of Nirenberg and Trèves [13] as well as those of Beals and Fefferman [1].

Theorem 8.2 was proved by Duistermaat and Hörmander [2] under the assumption that at every characteristic point the Hamilton fields of Re p and Im p are linearly independent of the radial vector field $\partial/\partial\xi$. Condition (P) implies that $\{\mathrm{Re}\, p, \mathrm{Im}\, p\} = 0$ on the characteristic set so it must then be an involutive manifold of codimension 2. Thus it has a natural foliation by two dimensional leaves B with tangent plane spanned by $H_{\mathrm{Re}\, p}$ and $H_{\mathrm{Im}\, p}$. Now recall the notation $s_u^*(x,\xi)$ for the number of L^2 derivatives of u at (x,ξ). It was proved in [2] that if $s_{Pu}^* \geqq s$ then $\min(s_u^*, s+m-1)$ is a superharmonic function with respect to the analytic structure defined in B by the Hamilton field H_p. Now suppose that u has the properties in (8.1). The lower semi-continuous function s_u^* must assume its infimum, and this must occur at a characteristic point (x,ξ) if it is $< s+m-1$, for $s_u^* \geqq s+m$ at the non-characteristic points. But then the superharmonic function $\min(s_u^*, s+m-1)$ must be constant in the leaf through (x,ξ) which is a contradiction with the assumption which states that it contains points over $\complement K$.

There is another situation where the following improvement of Theorem 6.2 will give Theorem 8.2:

THEOREM 8.3. *Let* P *be a pseudo-differential operator of order* 1 *with principal symbol* p, *and let* $\mathbf{R} \supset I \ni t \to \gamma(t) \in T^*(\Omega)\setminus 0$ *be a finite interval on a bicharacteristic for* Re p. *Assume that* $\mathrm{Im}\, p \geqq 0$ *in a neighborhood of* $\gamma(I)$. *If* $u \in \mathcal{D}'(\Omega)$, $Pu \in H_{(s)}$ *at* $\gamma(I)$ *and* $u \in H_{(s)}$ *at* $\gamma(b)$ *when* b *is the right hand end point of* I, *then* $u \in H_{(s)}$ *at* $\gamma(I)$.

The proof was given in Hörmander [10]. Let $\mathcal{C}$ be the characteristic set of p and let $\mathcal{C}_{11}$ be the set of points lying on some semibicharacteristic γ with a non-characteristic end point. From Condition (P) it follows easily that P must then satisfy the hypothesis of Theorem (8.3) at γ after multiplication by an elliptic factor, so (8.1) follows microlocally at $\mathcal{C}_{11}$. Note that a semibicharacteristic starting at a point in $\mathcal{C} \setminus \mathcal{C}_{11}$ must always remain in the characteristic set.

The preceding arguments must be extended to give a general proof of Theorem 8.2. An obvious difficulty is of course that as noted already we sometimes have propagation of singularities along curves and sometimes propagation along two dimensional manifolds. Fortunately these two cases can be nicely separated. Let us denote by $\mathcal{C}_2$ the set of characteristic points where $d \operatorname{Re} p$ and $d \operatorname{Im} p$ are linearly independent. As already observed this is an involutive manifold of codimension 2. Now we extend $\mathcal{C}_2$ to the set $\mathcal{C}_2^e$ of all points which can be joined by a semibicharacteristic to some point in $\mathcal{C}_2$. Note that a semibicharacteristic through a point in $\mathcal{C} \setminus (\mathcal{C}_{11} \cup \mathcal{C}_2^e)$ must be a *one dimensional bicharacteristic*, that is, a curve such that H_p is always proportional to the tangent vector along it.

PROPOSITION 8.4. *Condition* (P) *implies that* $\mathcal{C}_2^e$ *is an involutive manifold of codimension* 2.

To see this we let V be a small neighborhood of a point in $\mathcal{C}_2$. Then $V \cap \mathcal{C}_2$ divides the hypersurface of zeros of $\operatorname{Re} qp$ in V into two parts where $\operatorname{Im} qp > 0$ and $\operatorname{Im} qp > 0$ respectively. The flowout along $H_{\operatorname{Re} qp}$ of these sets must lie in the sets where $\operatorname{Im} qp \geqq 0$ and $\operatorname{Im} qp \leqq 0$ respectively, so the flowout of $V \cap \mathcal{C}_2$ remains in $\mathcal{C}$ and is involutive. (This shows that $\mathcal{C}_2^e \cap \mathcal{C}_{11} = \emptyset$.) It is not hard to complete the proof of Proposition 8.4.

As in the non-degenerate case $\mathcal{C}_2^e$ has a natural foliation by two dimensional leaves, which we call two dimensional bicharacteristics. From a point in $\mathcal{C}_2^e \setminus \mathcal{C}_2$ we can always find a semibicharacteristic γ to a point in $\mathcal{C}_2$ and it must be a one dimensional bicharacteristic until it hits the boundary of $\mathcal{C}_2$. It is then a routine matter to show that one can transform P by conjugation with Fourier integral operators as in section 6 so that γ becomes $I \times 0 \times \xi^0$, where I is an interval on the x_1 axis, $\xi^0 = (0, \cdots, 0, 1)$ and

$$p(x, \xi) = \xi_1 + if(x, \xi'), \ \xi' = (\xi_2, \cdots, \xi_n) .$$

Since $d \operatorname{Im} p \neq 0$ at some point on I it follows from condition (P) that

$$f(x, \xi') = g(x', \xi')\, h(x, \xi')$$

where $g = 0 \neq dg$ at $(0, \xi^0)$, g is homogeneous of degree 1, h is homogeneous of degree 0, and $h \geqq 0$. If we allow ourselves to make a non-homogeneous canonical transformation we can take $g = \xi_2$ and obtain

$$H_p = \partial/\partial x_1 + ih(x_1, x_2, 0, \xi^0)\partial/\partial x_2 .$$

As in the regular case we shall use H_p to define a natural complex structure in the leaves of the foliation, which of course are the parallels of the $x_1 x_2$ plane with these coordinates.

PROPOSITION 8.5. *Let* $a(x_1, x_2, y)$ *be a non-negative* C^∞ *function defined in a neighborhood of* $\{(x_1, 0, 0), x_1 \in I\} \subset \mathbf{R}^{2+k}$, *where* I *is a compact interval on* $\mathbf{R}$. *Assume that*

$$a(x_1, 0, 0) = 0, \ x_1 \in I .$$

Then there is an open neighborhood Ω *of* $I \times 0$ *in* $\mathbf{R}^{2+k}$ *such that the equation*

(8.2) $$\partial u/\partial x_1 + i\, a(x, y)\partial u/\partial x_2 = f$$

has a solution $u \in C^\infty(\Omega)$ *for every* $f \in C^\infty(\Omega)$.

This is easily proved since the estimates required are consequences of Theorem 8.3. A simple corollary is that if $u_0(x)$ is a solution of

$$\partial u_0/\partial x_1 + i\,a(x,0)\,\partial u_0/\partial x_2 = 0 \tag{8.3}$$

which is in C^∞ near $I\times 0$ in $\mathbf{R}^2$, then there is a C^∞ solution $u(x,y)$ of (8.2) in a neighborhood of $I\times 0$ in $\mathbf{R}^{2+k}$ which is equal to u_0 when $y=0$. It is also easy to show that one can find u_0 so that $\partial u_0/\partial x_2 \neq 0$ on I.

Let us now for a moment drop the parameters y and assume that I is a maximal interval such that $a(x_1,0)=0$ for $x_1 \in I$. Choose a closed rectangle R with axes parallel to the coordinate axes so that $u \in C^\infty(R)$, $I\times 0$ is in the interior of R and $a\neq 0$ on the sides parallel to the x_2 axis. We choose R so small that the argument of $\partial u/\partial x_2$ changes by less than $\pi/4$ in R. (Differentiation of (8.3) with respect to x_2 shows that $\partial u_0/\partial x_2$ is constant on $I\times 0$.) Then it is clear that intervals in R parallel to the x_1 or x_2 axis are mapped to C^1 curves with tangent direction differing by less than $\pi/4$ from $-i\,\partial u/\partial x_2$ resp. $\partial u/\partial x_2$, evaluated at a point on I. If we join two points in R by a curve consisting of two line segments parallel to the coordinate axes, it follows that the points have different images under u unless they lie on a line $x_2 =$ constant and a vanishes between them. In particular, the boundary of R is mapped to a Jordan curve Γ, which has $u(I)$ in its interior, and it is clear that $u(R)$ is equal to the interior of Γ together with Γ. It follows that u is a homeomorphism from $\tilde{R}$ to this subset of $\mathbf{C}$ if $\tilde{R}$ is obtained from R by identifying points in R which lie on a line segment $x_2 =$ constant where a vanishes identically.

Now assume that v is any other solution of (8.3) in a neighborhood of R. Then $v=f(u)$ where f is a continuous function from $u(R)$ to $v(R)$, and it is in fact analytic. This is obvious if $a\neq 0$ everywhere, for u is then a diffeomorphism which transforms (8.3) to the Cauchy-Riemann equation. In general we can obtain the same conclusion by extending u to a solution of the equation

$$\partial u/\partial x_1 + i(a(x)+\varepsilon^2)\partial u/\partial x_2 = 0$$

which for $\varepsilon = 0$ is the given solution. If v is extended similarly we obtain

$$v(x, \varepsilon) = f_\varepsilon(u(x, \varepsilon)), \ x \in R, \ \varepsilon \neq 0$$

where f_ε is analytic. When $\varepsilon \to 0$ we obtain $f_\varepsilon \to f$ so f is analytic.

Now let B_0 be the set of points in a two dimensional bicharacteristic which lie on a semibicharacteristic with both end points in $\mathcal{C}_2$, and let $\tilde{B}_0$ be the set obtained by identifying such points on the same embedded one dimensional bicharacteristic. Then we have just seen that the solutions of the equation $H_p w = 0$ define a natural structure of Riemann surface in $\tilde{B}_0$; the local analytic functions on $\tilde{B}_0$ lifted to B_0 are thus precisely the solutions of the equation $H_p w = 0$.

We can now state an extension of the superharmonicity theorem mentioned above:

THEOREM 8.6. *Let* P *be a pseudo-differential operator in* Ω *satisfying condition* (P), *and denote the order by* m. *Let* $u \in \mathcal{D}'(\Omega)$, *let* B *be a two dimensional bicharacteristic of* P *and* s *a function defined in* $\tilde{B}_0$ *such that* $\pi^* s \leqq s^*_{Pu}$ *where* $\pi : B_0 \to \tilde{B}_0$ *is the projection. Let* $\tilde{s}_u$ *be the largest function on* $\tilde{B}_0$ *such that* $\pi^* \tilde{s}_u \leqq s^*_u$, *that is,* $\tilde{s}_u$ *is the minimum of* s^*_u *on each equivalence class. If* s *is superharmonic in* $\omega \subset \tilde{B}_0$ *it follows then that*

$$\min(\tilde{s}_u, s+m-1) \text{ is superharmonic in } \omega .$$

In the proof of this theorem the following characterization of $H_{(s)}$ spaces is convenient:

LEMMA 8.7. *Let* χ *and* ϕ *be functions in* $C_0^\infty(T^*\mathbf{R}^n \setminus 0)$ *and set*

$$q_t(x,\xi) = \chi(x,\xi/t)\, t^{\phi(x,\xi/t)} .$$

If $f \in \mathcal{E}'(\mathbf{R}^n)$ *and* $s_f^* > \operatorname{Re}\phi$ *in* $\operatorname{supp}\chi$, *then* $\|q_t(x,D)f\|_{L^2}$ *is bounded as* $t \to \infty$. *Conversely, if* $\|q_t(x,D)f\|$ *is bounded as* $t \to \infty$ *then*

$$s_f^*(x,\xi) \geqq \operatorname{Re}\phi(x,\xi) \text{ if } \chi(x,\xi) \neq 0 .$$

Let I be a complete embedded one dimensional bicharacteristic. An easy modification of the proofs of Nirenberg and Trèves shows that, if $m = 1$,

(8.4) $$\|u\| \leqq C(\|Pu\| + \|Ru\|),\ u \in C_0^\infty$$

where R is a pseudo-differential operator which is of order -1 in a conic neighborhood V of I. Let B be the leaf through I and ω an open neighborhood of I in $\tilde{B}_0$ such that the inverse image does not meet V. Let K be a compact set in ω and h a harmonic polynomial such that

$$h(w) < \min(\tilde{s}_u, s) \text{ on } \partial_{\tilde{B}_0} K$$

where w is the local analytic coordinate. The theorem will be proved if we show that the same inequality is valid in K. That $h(w) < s$ follows since s is superharmonic so what must be proved is that $h(w) < \tilde{s}_u$. Now w can be considered as a solution of $H_p w = 0$, and if $h = \operatorname{Re} F$ where F is an analytic polynomial then $F(w)$ satisfies the same equation. It can be extended to a function ϕ satisfying the equation in a neighborhood of the inverse image K′ of K, in $\mathcal{C}_2^e$, so that for a suitable $\chi \in C_0^\infty$ which is equal to 1 near K′

$$s_u^* > \operatorname{Re}\phi \text{ in } \operatorname{supp} d\chi,\quad s_f^* > \operatorname{Re}\phi \text{ in } \operatorname{supp}\chi .$$

An iterative argument allows one to assume also that $s_u^* > \operatorname{Re}\phi - \delta$ in $\operatorname{supp}\chi$ for some fixed but arbitrarily small $\delta > 0$. Now we apply (8.4) with u replaced by $q_t(x,D)u$ where q_t is defined in Lemma 8.7. The main difficulty is then to estimate $\|Pq_t u\|$ or just $\|[P, q_t]u\|$. The leading term becomes

$$-i(H_p\chi(x, \xi/t) + \chi(x, \xi/t) \log t\, H_p\phi(x, \xi/t))\, t^{\phi(x,\xi/t)} .$$

If $H_p\phi$ were 0, which one can achieve by the Cauchy-Kovalevsky theorem in the analytic case, then the proof would be finished. In general one can only make sure that $H_p\phi = 0$ on $\mathcal{C}_2^e$. To handle the second term then requires a preliminary estimate of $g(x', D')u$, where g is one of the factors in the factorization of $\operatorname{Im} p$. It can be obtained by a refinement of Theorem 8.3; the proof requires no new idea beyond a sharper lower bound for pseudo-differential operators: If $a \in S^m_{\rho,\delta}(\mathbf{R}^n)$ is compactly supported and non-negative, then

$$\operatorname{Re}(a(x, D)u, u) \geqq -C\,\|u\|_{(s)}^2, \quad u \in C_0^\infty ,$$

provided that $2s \geqq m - 6(\rho - \delta)/5$. In particular one can take $s = 0$, $m = 1$, $\rho = 1-\delta$ if $\delta \leqq 1/12$, and this allows one to complete the preceding proof gaining $1/12$ of a derivative in each step.

Let us observe that in Theorem 8.6 the radial vector field is allowed to be a tangent of the two dimensional bicharacteristic B. If it is a tangent at one point then it is everywhere a tangent, so the two dimensional bicharacteristic is really a curve in the cosphere bundle $S^*(\Omega)$ and it has a natural affine structure. Superharmonicity is replaced by concavity on this curve.

The "method of descent" allows one to derive from Theorem 8.6 results on the propagation of singularities along certain one dimensional bicharacteristics. In fact, assume that the principal symbol is of the form

$$\xi_1 + ih(x, \xi')\, g(x', \xi')$$

near $I \times \xi^0$, where I is an interval on the x_1 axis, and $h \geqq 0$ is not identically 0 there. Then

$$\xi_1 + i\,h(x, \xi')(g(x', \xi') + \xi_{n+1})$$

also satisfies condition (P) and there is now a two dimensional bicharacteristic through $I \times \xi^0$. A solution of $Pu = f$ in $\mathbf{R}^n$ can be considered as a distribution in $\mathbf{R}^{n+1}$ which is independent of x_{n+1} and Theorem 8.6 implies an analogous result on the regularity on I where superharmonicity is replaced by concavity with respect to the parameter $\int h(x_1, 0, \xi'^0)\, dx_1$ on I. (Intervals where this is constant should be identified to a point.)

LEMMA 8.8. *Suppose that* $\xi_1 + if(x,\xi')$ *satisfies condition* (P) *near* $I \times \xi^{0'}$ *where* I *is an interval on the* x_1 *axis and* $\xi^0 = (0, \cdots, 0, 1)$, *assume that* $df = 0$ *there but that the Hessian is not* 0 *at* $(0, \xi^0)$ *say. In a neighborhood of* $I \times \xi^{0'}$ *we can then write*

$$f(x,\xi') = g(x',\xi')\, h(x,\xi') + r(x,\xi')$$

where $g(x',\xi') = f(0,x',\xi')$, h *and* r *are in* C^∞, r *is homogeneous of degree* 1 *and* $h \geqq 0$. *Moreover,* $r \equiv 0$ *unless the Hessian of* g *is positive (negative) semidefinite at* 0 *and then* r *must still vanish when* $g < 0$ (resp. $g > 0$).

The proof follows from Malgrange's preparation theorem. F. Trèves has given an example which shows that r cannot always be taken equal to 0.

When $r = 0$ we have just seen that Theorem 8.6 leads to a concavity statement for s_u^* on I. When r is not identically 0, the fact that r vanishes of infinite order on the set of sign changes combined with the improvement of Theorem 8.3 mentioned above, which permits one to get good control away from the set of sign changes, also leads to a slightly weaker result on the propagation of singularities along I.

Let us now return to the proof of (8.1). The results on propagation of singularities discussed above show that if u satisfies the hypothesis in (8.1) and $\inf s_u^* < s+m-1$, then the infimum cannot be attained at points of $\mathcal{C}_{11}$, $\mathcal{C}_2^e$ or the set $\mathcal{C}_{12} \subset \mathcal{C} \setminus (\mathcal{C}_{11} \cup \mathcal{C}_2^e)$ consisting of bicharacteristics of the form discussed in Lemma 8.8. What remains is the set $\mathcal{C}_3$ consisting of bicharacteristics where p can be put in the form $\xi_1 + if(x,\xi')$ with f vanishing to the third order. Now it turns out that the proofs of Beals and Fefferman give global $H_{(s)}$ estimates near this set, and they show that the infimum of s_u^* cannot be assumed in $\mathcal{C}_3$ only. The conclusion is that $s_u^* \geqq s+m-1$ which completes the proof. Detailed proofs are given in a paper to appear in Annals of Mathematics.

REFERENCES

[1] R. Beals and C. Fefferman, On local solvability of linear partial differential equations. Ann. of Math. 97(1973), 482-498.

[2] J. J. Duistermaat and L. Hörmander, Fourier integral operators II. Acta Math. 128(1972), 183-269.

[3] J. J. Duistermaat and J. Sjöstrand, A global construction for pseudo-differential operators with non-involutive characteristics. Inv. Mat. 20(1973), 209-225.

[4] Yu.V. Egorov, On canonical transformations of pseudo-differential operators. Uspehi Mat. Nauk 25(1969), 235-236.

[5] ———, Subelliptic operators. Usp. Mat. Nauk 30:2(1975), 57-114 (Russian Math. Surveys 30:2(1975), 59-118).

[6] L. Hörmander, Pseudo-differential operators and hypoelliptic equations. Proc. Symp. Pure Math. 10(1966), 138-183.

[7] ———, Fourier integral operators I. Acta Math. 127(1971), 79-183.

[8] ———, Linear differential operators. Actes Congr. Int. Math. 1970, 1, 121-133.

[9] ———, Uniqueness theorems and wave front sets for solutions of linear differential equations with analytic coefficients. Comm. Pure Appl. Math. 24(1971), 671-704.

[10] ———, On the existence and the regularity of solutions of linear pseudo-differential equations. Ens. Math. 17(1971), 99-163.

[11] R. Melrose, Equivalence of glancing hypersurfaces. Inv. Mat. 37 (1976), 165-191.

[12] S. Mizohata, Solutions nulles et solutions non analytiques. J. Math. Kyoto Univ. 1(1962), 271-302.

[13] L. Nirenberg and F. Trèves, On local solvability of linear partial differential equations. Part I: Necessary conditions. Comm. Pure Appl. Math. 23(1970), 1-38. Part II: Sufficient conditions. Comm. Pure Appl. Math. 23(1970), 459-510.

[14] M. Sato, T. Kawai and M. Kashiwara, Microfunctions and pseudo-differential equations. Lecture notes in math. 287(1973), 263-529.

[15] F. Trèves, A new method of proof of the subelliptic estimates. Comm. Pure Appl. Math. 24(1971), 71-115.

[16] K. Yamamoto, On the reduction of certain pseudo-differential operators with non-involutive characteristics. J. Diff. Eq. 26(1977), 435-442.

FOURIER INTEGRAL OPERATORS WITH COMPLEX PHASE FUNCTIONS

J. Sjöstrand[1)]

0. *Introduction*

In his lectures L. Hörmander introduced local Fourier integral operators in order to transform pseudo-differential operators. Another important application of Fourier integral operators is the construction of solutions to homogeneous pseudo-differential equations; $Pu \equiv 0 \mod C^\infty$. Such constructions sometimes also lead to parametrices (i.e. inverses modulo smoothing operators). In this context, it is often important to study Fourier integral operators globally as was done first by Hörmander [4] and Duistermaat-Hörmander [2]. If the leading symbol of the pseudo-differential equation under study takes complex values it is often necessary to consider complex valued phase functions. We shall here outline the calculus in the complex case. This calculus was developed jointly with A. Melin [10] and we refer to this paper for more details. A parallel theory has been developed by several Soviet mathematicians, see Maslov [8], Kucherenko [6, 7], Danilov-Maslov [1].

1. *Local study*

Let $\Omega \subset \mathbf{R}^n$ be open. By $S_C^m(\Omega \times \mathbf{R}^N) \subset S_{1,0}^m(\Omega \times \mathbf{R}^N)$ we denote the

---

1)Supported in part by NSF grant MCS77-18723.

Seminar on Singularities of Solutions
0-691-08221-9/78/0051 $00.70/1 (cloth)
0-691-08213-8/78/0051 $00.70/1 (paperback)

space of classical symbols $a(x,\theta)$ having an asymptotic expansion

$$(1.1)\quad a(x,\theta) \sim \sum_{j=0}^{\infty} a_{m-j}(x,\theta),\ \ a_{m-j}(x,\lambda\theta) = \lambda^{m-j} a_{m-j}(x,\theta),\ \ \lambda > 0 .$$

Let $V \subset \Omega \times \dot{\mathbf{R}}^N$ be an open cone, $\dot{\mathbf{R}}^N = \mathbf{R}^N \setminus \{0\}$. We say that $\phi(x,\theta) \in C^\infty(V)$ is a *non-degenerate phase function* if the following four conditions hold:

$$(1.2)\qquad \operatorname{Im} \phi(x,\theta) \geq 0$$

$$(1.3)\qquad \phi(x,\lambda\theta) = \lambda\phi(x,\theta),\ \ \lambda > 0$$

$$(1.4)\qquad d\phi \neq 0 \text{ everywhere}$$

(1.5) $d(\partial\phi/\partial\theta_1), \cdots, d(\partial\phi/\partial\theta_N)$ are linearly independent on the set $(C_\phi)_{\mathbf{R}} = \{(x,\theta) \in V;\ \phi'_\theta(x,\theta) = 0\}$.

If $a \in S^m_c$ has its support in some cone $W \subset V$ such that $\{(x,\theta) \in W;\ |\theta| = 1\}$ is compact, then we can define a *Fourier integral distribution* $I(a,\phi) \in \mathcal{D}'(\Omega)$ formally as

$$(1.6)\qquad I(a,\phi)(x) = \int e^{i\phi(x,\theta)} a(x,\theta)\, d\theta .$$

More precisely, if $u \in C_0^\infty(\Omega)$, we can give a sense to the oscillatory integral

$$(1.7)\qquad \langle I(a,\phi), u\rangle = \iint e^{i\phi(x,\theta)} a(x,\theta)\, u(x)\, dx d\theta$$

by making suitable formal partial integrations which decrease the order of the symbol a. If we define $(\Lambda_\phi)_{\mathbf{R}}$ as $\{(x,\phi'_x(x,\theta));\ (x,\theta) \in (C_\phi)_{\mathbf{R}}\}$, then it is not hard to show that

$$(1.8)\qquad \text{S.S.}\,(I(a,\phi)) \subset (\Lambda_\phi)_{\mathbf{R}} \quad (\text{S.S.} = \text{Singular Spectrum}) .$$

EXAMPLE. Let $f \in C^\infty(\Omega)$, $\operatorname{Im} f \geq 0$ and $df \neq 0$ where $f(x) = 0$. Then $\int_0^\infty e^{itf(x)}dt = \frac{i}{(f(x)+i0)}$ is a Fourier integral distribution.

The main difficulty in order to define Fourier integral distributions globally is that it is often difficult to find suitable phase functions globally. One therefore has to work with sums of expressions like $I(a,\phi)$. The main problem then turns out to be to examine which phase functions give rise to the same spaces of Fourier integral distributions microlocally.

2. *Lagrangean manifolds associated to phase functions*

Let $W \subset \mathbf{R}^n$ be open and let $f \in C^\infty(W)$. If $\tilde{W} \subset \mathbf{C}^n$ is open with $\tilde{W} \cap \mathbf{R}^n = W$, then it is well known that there exists a function $\tilde{f} \subset C^\infty(\tilde{W})$ such that $\bar{\partial}\tilde{f}$ vanishes to infinite order on W and $\tilde{f}|_W = f$. Moreover, $\tilde{f}$ is unique up to a function vanishing to infinite order on W. (One way of constructing $\tilde{f}$ is to write down what the Taylor series should be at the real points and then make it converge as in the classical Borel theorem.) We call $\tilde{f}$ the *almost analytic extension* of f. There is also a natural notion of almost analytic manifolds in the complex domain. It turns out that modulo certain natural equivalence relations, many important geometric objects, associated with complex phase functions, are almost analytic. This terminology is quite heavy however, so in order to simplify this survey we shall, somewhat incorrectly, think of all functions in the real domain as being real analytic. The corresponding extensions are then germs of holomorphic functions defined in a small neighborhood of the real domain. If W is an open set in the real domain, then $\tilde{W}$ will denote some sufficiently small open set in the complex domain which intersects the real domain along W. If $\Lambda \subset \tilde{W}$ is a complex manifold, we put $\Lambda_{\mathbf{R}} = W \cap \Lambda$.

Now let ϕ be a non-degenerate phase function defined in a conic neighborhood V of a point $(x_0, \theta_0) \in \Omega \times \dot{\mathbf{R}}^N$, and assume that $(x_0, \theta_0) \in (C_\phi)_{\mathbf{R}}$. We extend ϕ to some complex cone $\tilde{V}$ and put

(2.1) $$C_\phi = \{(x,\theta) \in \tilde{V};\ \phi'_\theta(x,\theta) = 0\},$$

if $\tilde{V}$ is small enough then C_ϕ is a manifold of complex dimension n. The map

(2.2) $$C_\phi \ni (x,\theta) \mapsto (x, \phi'_x) \in \widetilde{T^*\Omega \setminus 0}$$

has injective differential so if we shrink V and $\tilde{V}$ around (x_0, θ_0), the image $\Lambda = \Lambda_\phi$ will be a conic n-dimensional manifold. It is easily verified that Λ is Lagrangean. Let $(x_0, \xi_0) = (x_0, \phi'_x(x_0, \theta_0)) \in \Lambda_{\mathbf{R}}$. We say that Λ is *generated by the phase function* ϕ *near* (x_0, ξ_0).

After a real change of coordinates near x_0 we can assume that the projection $\Lambda \ni (x,\xi) \mapsto \xi$ has bijective differential, so that Λ takes the form $x = x(\xi)$ near (x_0, ξ_0) (cf. Hörmander's lecture notes). Then $x(\xi) = \partial H/\partial \xi$ if $H(\xi) = \langle x(\xi), \xi \rangle$. It is easily verified that $\det(\text{Hess}\, \phi) \neq 0$ with this choice of x-coordinates, and if $(x(\xi), \theta(\xi))$ is the inverse image of $(x(\xi), \xi)$ in (2.2), then $(x(\xi), \theta(\xi))$ is the non-degenerate critical point of the function $(x,\theta) \mapsto \phi(x,\theta) - \langle x, \xi \rangle$. In view of Euler's homogeneity relations, we have $\phi(x(\xi), \theta(\xi)) = 0$ so the corresponding critical value is $-H(\xi)$. We now need a lemma (that we formulate only in the real analytic case).

LEMMA 2.1. *Let* $a(y,\alpha)$ *be defined near* $(0,0)$ *in* $\mathbf{R}^n \times \mathbf{R}^k$. *Assume that* $\operatorname{Im} a(y,\alpha) \geq 0$ *with equality at* $(0,0)$, $a'_y(0,0) = 0$, $\det a''_{yy}(0,0) \neq 0$. *Let* $y(\alpha) \in \mathbf{C}^n$ *be the analytic solution of* $a'_y(y,\alpha) = 0$, $y(0) = 0$, *defined for* α *in a small real neighborhood of* 0. *Then there exists a constant* $C > 0$ *such that*

$$\operatorname{Im} a(y(\alpha),\alpha) \geq C|\operatorname{Im} y(\alpha)|^2 .$$

This lemma shows that $\operatorname{Im} H(\xi) \leq 0$ for real ξ in a neighborhood of ξ_0. One can also show easily that for some constant $C > 0$:

$$\frac{1}{C}\,|\operatorname{Im}(x,\theta)|^2 \leq |\operatorname{Im}(x,\phi'_x)| \leq C|\operatorname{Im}(x,\theta)|^{1/2}$$

for $(x, \theta) \in C_\phi$ in some small neighborhood of (x_0, θ_0), so that $(\Lambda_\phi)_R$ is the image of $(C_\phi)_R$. This estimate is particularly important in the C^∞-case, in order to verify that Λ_ϕ is almost analytic.

It is clear that $\Lambda = \Lambda_\phi$ is a positive Lagrangean manifold in the sense of the following definition.

DEFINITION 2.2. Let $\Lambda \subset \widetilde{T^*\Omega \setminus 0}$ be a conic Lagrangean manifold. We say that Λ is positive if for every point $(x_0, \xi_0) \in \Lambda_R$, we can choose real coordinates near x_0 such that Λ takes the form $x = \partial H/\partial \xi$ near (x_0, ξ_0), where H is homogeneous of degree 1 and satisfies $\operatorname{Im} H(\xi) \leq 0$ for real ξ.

Locally every positive Lagrangean manifold can be generated by a non-degenerate phase function. We just represent Λ by $x = \partial H/\partial \xi$ and choose $\phi(x, \xi) = \langle x, \xi \rangle - H(\xi)$.

3. *Global definition of Fourier integral distributions*

The main fact here is that two phase functions generate the same spaces of distributions microlocally if they generate the same Lagrangean manifolds. More precisely, we have

THEOREM 3.1. *Let $\phi(x, \theta)$, $\psi(x, w)$ be non-degenerate phase functions defined in some conic neighborhoods of $(x_0, \theta_0) \in \Omega \times \dot{R}^N$ and $(x_0, w_0) \in \Omega \times \dot{R}^M$ respectively such that $(x_0, \theta_0) \in C_\phi$, $(x_0, w_0) \in C_\psi$, $\phi'_x(x_0, \theta_0) = \psi'_x(x_0, w_0) = \xi_0$ and such that Λ_ϕ and Λ_ψ coincide near (x_0, ξ_0). Then for every $a \in S_C^{m+(n-2N)/4}(\Omega \times R^N)$ with support in a small but fixed conic neighborhood of (x_0, θ_0), there exists a symbol $b \in S_C^{m+(n-2M)/4}(\Omega \times R^M)$ with support in a small fixed neighborhood of (x_0, w_0) such that $I(a, \phi) \equiv I(b, \psi) \mod C^\infty$.*

Sketch of the proof. Choose local coordinates near x_0 such that $\Lambda_\phi = \Lambda_\psi$ takes the form $x = \partial H/\partial \xi$. Then by applying the complex version of the stationary phase method [9], we get an asymptotic formula for the Fourier transform of $I(a,\phi)$ in a conic neighborhood of ξ_0.

$$\widehat{I(a,\phi)(\xi)} \sim e^{-iH(\xi)}A(\xi), \quad \xi \to \infty, \tag{3.1}$$

where $A \in S_c^{m-n/4}$ and has the leading homogeneous term

$$\left[\frac{a_{m+(n-2N)/4}}{\left(\det \frac{1}{(2\pi i)} \text{Hess } \phi\right)^{1/2}}\right](x(\xi), \theta(\xi)).$$

Here $a_{m+(n-2N)/4}$ is the leading homogeneous term of a and we make the natural choice of the square root of the determinant for non-degenerate symmetric matrices with positive semidefinite real part. Conversely, if $A \in S_c^{m-n/4}$ is given, we can construct $a \in S_c^{m+(n-2N)/4}$ by iteration so that (3.1) is valid. Applying this converse argument to $\widehat{I(b,\psi)}$, we can then construct the desired symbol b.

DEFINITION 3.2. Let $\Lambda \subset \widetilde{T^*\Omega \setminus 0}$ be a positive conic Lagrangean manifold such that Λ_R is closed. Then $I_c^m(\Omega,\Lambda)$ is the space of all $u \in \mathcal{D}'(\Omega)$ such that

1° $S.S.(u) \subset \Lambda_R$.

2° If $(x_0, \xi_0) \in \Lambda_R$ and $\phi(x,\theta)$ is a non-degenerate phase function defined in a conic neighborhood of $(x_0, \theta_0) \in \Omega \times \dot{R}^N$, $(x_0, \theta_0) \in C_\phi$, $\xi_0 = \phi'_x(x_0, \theta_0)$, which generates Λ near (x_0, ξ_0), then for some $a \in S_c^{m+(n-2N)/4}(\Omega \times R^N)$ with support in a small conic neighborhood of (x_0, θ_0), we have $(x_0, \xi_0) \notin S.S.(u - I(a,\phi))$.

The stationary phase method shows that classical pseudo-differential operators of order 0 preserve the spaces $I_c^m(\Omega,\Lambda)$. Then by a pseudo-differential partition of unity it is easy to see that every $u \in I_c^m(\Omega,\Lambda)$ can be written as a locally finite sum of different $I(a,\phi)$'s where ϕ generates Λ locally.

REMARK 3.3. In the proof of Theorem 3.1, we get a relation $a_{m+(n-2N)/4} = \mathcal{H}_{\phi,\psi} b_{m+(n-2M)/4}$ between the leading homogeneous symbols evaluated at corresponding points of C_ϕ and C_ψ. If we let Ω be a manifold and define $u \in I_C^m$ rather as a distribution density of order ½, then we may use the above relations to define the principal symbol of u, which is a section of a certain complex (trivial) line bundle over Λ. Roughly, one can say that the principal symbol is a ½-density but we shall not go into the details in this survey.

4. *Fourier integral operators*

Let $X \subset R^{n_1}$, $Y \subset R^{n_2}$, $Z \subset R^{n_3}$ be open. We say that $C \subset \widetilde{T^*(X \times Y)\setminus 0}$ is a positive canonical relation if C_R is closed and contained in $(T^*X\setminus 0) \times (T^*Y\setminus 0)$ and $C' = \{(x,\xi,y,-\eta);\ (x,\xi,y,\eta) \in C\}$ is a (conic) positive Lagrangean manifold. If $K_A \in I_C^m(X \times Y, C')$ and $A: C_0^\infty(Y) \to C^\infty(X)$ is the operator with distribution kernel K_A, we call A a Fourier integral operator. A special case is obtained when C is the graph of a canonical transformation. Under certain assumptions, the composition of two Fourier integral operators is again a Fourier integral operator.

THEOREM 4.1. *Let* $\Delta_Y = T^*X \times \operatorname{diag}(T^*Y \times T^*Y) \times T^*Z$, *and let* $C_1 \subset \widetilde{T^*(X \times Y)\setminus 0}$, $C_2 \subset \widetilde{T^*(Y \times Z)\setminus 0}$ *be positive canonical relations. We assume*

(4.1) *The natural projection*

$$C_1 \times C_2 \cap \Delta_Y \to (T^*X\setminus 0) \times (T^*Z\setminus 0)$$

is injective and proper.

(4.2) $C_1 \times C_2$ *and* $\tilde{\Delta}_Y$ *intersect transversally at every point in* $C_1 \times C_2 \cap \Delta_Y$.

Then $C_1 \circ C_2$ *is a well-defined positive canonical relation. Moreover, if*

$K_A \in I_C^{m_1}(X \times Y, C_1')$, $K_B \in I_C^{m_2}(Y \times Z, C_2')$ *and* A *or* B *is properly supported, then* $K_{A \circ B} \in I_C^{m_1+m_2}(X \times Z, (C_1 \circ C_2)')$.

The essential idea of the proof is to write $K_{A \circ B}(x, z) = \int K_A(x, y) \cdot K_B(y, z)\, dy$ and introduce microlocal explicit expressions for K_A and K_B. If the corresponding phase functions are $\phi(x, y, \theta)$ and $\psi(y, z, w)$ and we consider (θ, w, y) as fiber variables we can say that $K_{A \circ B}(x, z)$ is given by an oscillatory integral with phase $\phi(x, y, \theta) + \psi(y, z, w)$.

We notice finally that the calculus we have developed also works on manifolds without any modifications.

5. *Application to the exponential of a pseudo-differential operator*

For the applications it is important to have a simpler and more geometric definition of the positivity of Lagrangean manifolds. On $\widetilde{T^*\Omega \setminus 0}$ we put

$$S(x, \xi) = -\langle \operatorname{Im} x, \operatorname{Re} \xi \rangle . \tag{5.1}$$

If $\omega = \sum \xi_j dx_j$ is the symplectic 1-form, we can consider $S(x, \xi)$ as the value of $-\omega$ applied to the tangent vector $(\operatorname{Im} x, \operatorname{Im} \xi)$ at the point $(\operatorname{Re} x, \operatorname{Re} \xi)$. It is then easy to show that the function S is invariant under (extensions of) real homogeneous canonical transformations, up to an error which is locally $O(|(\operatorname{Im} x, \operatorname{Im} \xi)|^3)$.

LEMMA 5.1. *Let* $\Lambda \subset \widetilde{T^*\Omega \setminus 0}$ *be a conic Lagrangean manifold. Then* Λ *is positive iff for every real point* $(x_0, \xi_0) \in \Lambda_R$ *there is a constant* $C > 0$ *such that*

$$S(x, \xi) \geq -C|\operatorname{Im}(x, \xi)|^3, \quad (x, \xi) \in \Lambda \tag{5.2}$$

in some neighborhood of (x_0, ξ_0).

The proof is not difficult, once we have observed that if Λ has the form $x = \partial H / \partial \xi$, with H positively homogeneous of degree 1, then for

$(x, \xi) = (\partial H/\partial \xi, \xi) \in \Lambda$:

$$\langle -\mathrm{Im}\, x, \mathrm{Re}\, \xi \rangle = -\mathrm{Im}\, H(\mathrm{Re}\, \xi) - \frac{1}{2} \langle \mathrm{Im}\, H''(\mathrm{Re}\, \xi)\, \mathrm{Im}\, \xi, \mathrm{Im}\, \xi \rangle + O(|\mathrm{Im}\, \xi|^3)$$

locally uniformly.

From now on we assume for simplicity that Ω is a compact manifold and let $P \in L^1_c(\Omega)$ be a classical pseudo-differential operator with principal symbol p. We want to construct a smooth family of Fourier integral operators $A_t : C^\infty(\Omega) \to C^\infty(\Omega)$, $t \geq 0$ so that

(5.3) $$A_0 = I$$

(5.4) The distribution kernel of $(D_t + P)A_t$ is a smooth function on

$$\overline{R}_+ \times \Omega \times \Omega \quad \left(D_t = \frac{1}{i}\frac{d}{dt}\right).$$

Roughly we are interested in $\exp(-itP)$. A very reasonable requirement for this problem to be well posed is that

(5.5) $$\mathrm{Im}\, p \leq 0 \quad \text{(on the real domain)}$$

so we assume (5.5) from now on. In the case when p is real valued, (5.3)-(5.4) is a simple hyperbolic problem and the solution has been known for a long time. In particular, it is known that the canonical relations of A_t should be the graph of $\exp t\mathcal{H}_p$ where $\mathcal{H}_p = \sum \frac{\partial p}{\partial \xi_j}\frac{\partial}{\partial x_j} - \frac{\partial p}{\partial x_j}\frac{\partial}{\partial \xi_j}$ is the Hamilton field of p. (The principal symbol and the lower order symbols of A_t are then determined by solving certain transport equations, which are essentially first order ordinary differential equations in t.)

It turns out that the same procedure works in the complex case, if we let $\mathcal{H}_p$ denote the unique real vector field in the complex domain such that

(5.6) $$\mathcal{H}_p(u) = \sum \frac{\partial p}{\partial \xi_j}\frac{\partial u}{\partial x_j} - \frac{\partial p}{\partial x_j}\frac{\partial u}{\partial \xi_j}$$

for all holomorphic functions u. The essential difficulty is then to show that $C_t = \text{graph}(\exp t\mathcal{H}_p)$ are well defined positive canonical relations for $t \geq 0$, and we shall here only outline how this particular difficulty can be overcome, following [11]. (Other treatments of essentially the same problem have been given by Kucherenko [6], Maslov [8], Danilov-Maslov [1], and a local solution of our problem would also follow from Trèves [13].)

We know that the positivity of C_t is measured by the function $S(x,\xi) - S(y,\eta)$ of $(x,\xi,y,\eta) \in C_t$, so we shall study how S develops along an integral curve of $\mathcal{H}_p$.

LEMMA 5.2. *With* $\rho = (x,\xi)$, *we have locally uniformly*

$$\mathcal{H}_p(S)(\rho) = -\text{Im}\, p(\text{Re}\,\rho) - \frac{1}{2} \langle \text{Im}\, p''(\text{Re}\,\rho)\,\text{Im}\,\rho,\ \text{Im}\,\rho \rangle + O(|\text{Im}\,\rho|^3 + |\text{Im}\,\rho|^2 |\text{Im}\, p'(\text{Re}\,\rho)|)\ . \tag{5.6}$$

This lemma is proved, using Taylor expansions and the Euler homogeneity relations. Using standard inequalities for non-negative functions we obtain from (5.6) that

$$\mathcal{H}_p(S)(\rho) \geq -\frac{1}{2}\,\text{Im}\, p(\text{Re}\,\rho) + O(|\text{Im}\,\rho|^3) \tag{5.7}$$

locally uniformly.

Now let $[0,T] \ni t \mapsto \gamma(t)$ be a real integral curve of $\mathcal{H}_p$, where $0 \leq T < +\infty$. We shall study $\exp t\mathcal{H}_p(y,\eta)$ for (y,η) close to $\gamma(0)$ and $t \in [0, T+\varepsilon]$ where $\varepsilon > 0$ is some small number. Let $\rho_t = \exp t\mathcal{H}_p(y,\eta)$, $0 \leq t \leq T+\varepsilon$. In the following C will denote some positive constant different from time to time. From (5.7) we get

$$\frac{d}{dt} S(\rho_t) \geq -\frac{1}{2}\,\text{Im}\, p(\text{Re}\,\rho_t) - C|\text{Im}\,\rho_t|^3\ . \tag{5.8}$$

In order to apply this estimate we need to estimate the variation of $|\text{Im}\,\rho_t|$. We have by Taylor expansion at $\text{Re}\,\rho_t$

$$\frac{d}{dt} \operatorname{Im} \rho_t = (\operatorname{Im} \mathcal{H}_p)(\rho_t) = A(t) \operatorname{Im} \rho_t + \operatorname{Im} \mathcal{H}_p(\operatorname{Re} \rho_t)$$

where $A(t)$ is a bounded matrix (depending also on ρ_t). Regarding this as a linear equation for $\operatorname{Im} \rho_t$ we get

$$(5.9) \qquad |\operatorname{Im} \rho_s| \leq C\left(|\operatorname{Im} \rho_t| + \left|\int_s^t |\operatorname{Im} p'(\operatorname{Re} \rho_\tau)|\, d\tau\right|\right),$$

$$0 \leq s,\ t \leq T+\varepsilon .$$

But $|\operatorname{Im} p'(\operatorname{Re} \rho_\tau)| \leq C(-\operatorname{Im} p(\operatorname{Re} \rho_\tau))^{1/2}$, so it follows that

$$(5.10) \qquad |\operatorname{Im} \rho_s| \leq C\left(|\operatorname{Im} \rho_t| + \left|\int_s^t - \operatorname{Im} p(\operatorname{Re} \rho_\tau)\, d\tau\right|^{1/2}\right),$$

$$0 \leq s,\ t \leq T+\varepsilon .$$

Integrating (5.8) we obtain

$$(5.11) \qquad S(\rho_t) - S(\rho_s) \geq -\frac{1}{2}\int_s^t \operatorname{Im} p(\operatorname{Re} \rho_\sigma)\, d\sigma$$

$$- C\int_s^t |\operatorname{Im} \rho_\sigma|^3\, d\sigma, \quad 0 \leq s \leq t \leq T+\varepsilon .$$

Using (5.10) to estimate $\operatorname{Im} \rho_\sigma$ and the fact that $\int_s^t - \operatorname{Im} p(\operatorname{Re} \rho_\sigma)\, d\sigma$ is as small as we like after decreasing ε and the neighborhood of $\gamma(0)$ where $\rho_0 = (y, \eta)$ belongs, we conclude that

$$(5.12) \qquad S(\rho_t) - S(\rho_s) \geq -\frac{1}{3}\int_s^t \operatorname{Im} p(\operatorname{Re} \rho_\sigma)\, d\sigma - C|\operatorname{Im} \rho_t|^3$$

when $0 \leq s \leq t \leq T+\varepsilon$.

Assuming for a moment that $\operatorname{Im} \rho_0 = 0$, we get from (5.12):

$$S(\rho_t) \geq -C|\mathrm{Im}\,\rho_t|^3 . \tag{5.13}$$

Taking squares of (5.10) and using (5.12) to estimate $-\int_s^t \mathrm{Im}\,p(\mathrm{Re}\,\rho_\sigma)\,d\sigma$ we get for $0 \leq s \leq t \leq T+\varepsilon$

$$|\mathrm{Im}\,\rho_s|^2 \leq C_1(|\mathrm{Im}\,\rho_t|^2 + 3(S(\rho_t) - S(\rho_s)) + C_2|\mathrm{Im}\,\rho_t|^3) , \tag{5.14}$$

or rather

$$|\mathrm{Im}\,\rho_s|^2 + 3C_1 S(\rho_s) \leq C_3|\mathrm{Im}\,\rho_t|^2 + 3C_1 S(\rho_t) . \tag{5.15}$$

Using (5.13) with t replaced by s and estimating $S(\rho_t)$ by a constant times $|\mathrm{Im}\,\rho_t|$ we get

$$|\mathrm{Im}\,\rho_s|^2 \leq C|\mathrm{Im}\,\rho_t| , \quad 0 \leq s \leq t \leq T+\varepsilon . \tag{5.16}$$

Dropping the assumption that $\mathrm{Im}\,\rho_0 = 0$, we then get more generally:

$$|\mathrm{Im}\,\rho_s|^2 \leq C(|\mathrm{Im}\,\rho_0| + |\mathrm{Im}\,\rho_t|), \quad 0 \leq s \leq t \leq T+\varepsilon . \tag{5.17}$$

This inequality means in particular that an integral curve of $\mathcal{H}_p$ cannot start at a real point, then leave the real domain and come back afterwards. In fact, (5.17) can be used to show that $C_t = \mathrm{graph}\,\exp t\mathcal{H}_p$ is a well defined canonical relation near $(\gamma(t), \gamma(0))$ and the positivity is provided by (5.12) (with $s = 0$). It is not difficult to deduce the following global result from our inequalities:

PROPOSITION 5.3. $C_t = \mathrm{graph}(\exp t\mathcal{H}_p)$, $0 \leq t < \infty$ *are well defined positive canonical relations with* $(C_t)_{\mathbf{R}} = \{((\exp t\mathcal{H}_p)(y,\eta), (y,\eta));$ $(\exp s\mathcal{H}_p)(y,\eta)$ *is real for* $0 \leq s \leq t\}$.

Studying also the transport equations one obtains

THEOREM 5.4. *There exists a smooth family* $A_t \in I_c^0(\Omega\times\Omega; C_t')$ *satisfying* (5.3), (5.4).

It is also easy to see from the transport equations that A_t are unique up to smoothing operators. In some cases it is possible to use $A_t \approx \exp(-itP)$ to obtain parametrices (i.e. inverses modulo smoothing operators). For instance, we have (still under the assumption (5.5)):

THEOREM 5.5. *Assume that there is a* $T > 0$ *such that* $\{(y,\eta) \in T^*\Omega\backslash 0;\ p(y,\eta) = 0, (\exp s\mathcal{H}_p)(y,\eta)$ *is real for* $0 \leq s \leq T\} = \emptyset$. *Then* P *has a twosided parametrix* E, *with*

$$\begin{aligned} S.S.'(E) = {} & \operatorname{diag}((T^*\Omega\backslash 0) \times (T^*\Omega\backslash 0)) \cup \\ & \{(\exp t\mathcal{H}_p(y,\eta), (y,\eta)); p(y,\eta) = 0, \exp s\mathcal{H}_p(y,\eta) \text{ is real for } 0 \leq s \leq t\}. \end{aligned}$$

The additional assumption here means that $(C_t)_R$ will be empty near $p^{-1}(0) \times p^{-1}(0)$ when t is sufficiently large. Then it is not difficult to give a meaning to the integral

$$E = i \int_0^\infty A_t dt . \tag{5.18}$$

We have

$$PE = i \int_0^\infty PA_t dt \equiv -i \int_0^\infty D_t A_t dt \equiv -\int_0^\infty \frac{d}{dt} A_t dt \equiv A_0 \equiv I$$

modulo smoothing operators, so E is a right parametrix. If $B_t = [P, A_t] \in I_c^1(\Omega \times \Omega, C_t')$, we have $B_0 = 0$ and $(D_t + P) B_t \equiv 0$ modulo smoothing operators. Looking at the corresponding transport equations we then conclude that $B_t \equiv 0$ and it follows that $EP \equiv PE \equiv I$.

Another interesting case is when P has double characteristics. Then the singularities of A_t may be stationary and the definition of the integral (5.18) is more troublesome. However, with suitable conditions on the subprincipal symbol of P one may hope that A_t decreases exponentially fast in some sense as $t \to \infty$. Results in this direction have been announced by Kucherenko [7]. Other results have been obtained by Helffer [3] and Menikoff-Sjöstrand [12]. Helffer uses Melin's inequality [9], to show that the L^2-norm of the operator A_t may decrease exponentially in some cases. Conversely, if one had the optimal results about the L^2-norm of operators with complex phase, then probably Melin's inequality would follow from these results applied to A_t. Notice also that Hörmander [5] used the sharp Gårding inequality (which is weaker than Melin's inequality) to obtain the behavior of the singular spectrum of solutions to $Pu = 0$, under the assumptions of Theorem 5.5.

REFERENCES

[1] V. G. Danilov and V. P. Maslov, Quasi-invertibility of functions of ordered operators in the theory of pseudo-differential operators, Journal of Soviet Mathematics 7 (1977), no. 5, 695-794.

[2] J. J. Duistermaat and L. Hörmander, Fourier integral operators II, Acta Math. 128 (1971), 183-269.

[3] B. Helffer, Quelques exemples d'opérateurs pseudo-differentiels localement resolubles, Proceedings from the meeting in PDE at St-Jean-de-Monts (1977), Springer Lecture notes (to appear).

[4] L. Hörmander, Fourier integral operators I, Acta Math. 127 (1971), 79-183.

[5] ——————, On the existence and regularity of solutions of linear pseudo-differential equations, L'Eneign. Math. 17 (1971), 99-163.

[6] V. V. Kucherenko, Asymptotic solutions of equations with complex characteristics, Mat. Sb. 95 (137) (1974), 164-213. Math. USSR Sb. 24 (1974), no. 2.

[7] ——————, Parametrix for equations with degenerate symbol, Sov. Math. Dokl., 17 (1976), no. 4.

[8] V. P. Maslov, Operational methods, MIR Publishers, Moscow (1976), (Russian edition in 1973).

[9] A. Melin, Lower bounds for pseudo-differential operators, Ark. Mat. 9, (1971), 117-140.

[10] A. Melin and J. Sjöstrand, Fourier integral operators with complex valued phase functions, Springer Lecture Notes 459, 120-223.

[11] ——————, Fourier integral operators with complex phase functions and parametrix for an interior boundary value problem, Comm. in PDE, 1 (1976), no. 4, 313-400.

[12] A. Menikoff and J. Sjöstrand, On the eigenvalues of a class of hypoelliptic operators (to appear, see also Menikoff's lecture notes to this seminar).

[13]. F. Trèves, Solutions of Cauchy problems modulo flat functions, Comm. in PDE, 1 (1976), no. 1.

HYPOELLIPTIC OPERATORS WITH DOUBLE CHARACTERISTICS

A. Menikoff[1)]

1. *Conditions for hypoellipticity*

Let $P(x, D)$ be a pseudo-differential operator of order m in $\Omega \subset \mathbf{R}^n$. We shall say that P is hypoelliptic if

$$\text{sing supp } u = \text{sing supp } Pu, \; u \in \mathcal{D}'(\Omega) .$$

Further we shall say that P is *hypoelliptic with loss of* μ *derivatives* if $u \in \mathcal{D}'(\Omega)$, $Pu \in H^{loc}_{(s)}$ implies $u \in H^{loc}_{(s+m-\mu)}$. In this definition μ measures the deviation from the ideal case of an elliptic operator which is hypoelliptic with no loss of derivatives. An easy application of the closed graph theorem gives

LEMMA 1.1. *If* P *is hypoelliptic with loss of* μ *derivatives, then for any compact set* $K \subset \Omega$ *and* $s, s' \in \mathbf{R}$ *there is a constant* C *such that*

$$\|u\|_{s+m-\mu} \leqq C(\|Pu\|_s + \|u\|_{s'}), \; u \in C_0^\infty(K) . \tag{1.1}$$

When $\mu < 1$ the converse is true and P is called subelliptic. Such operators will be discussed later in the seminar. In general the converse of Lemma 1.1 is false but it will also be valid in the case studied here.

---

1) Supported in part by NSF grant MCS77-18723.

Seminar on Singularities of Solutions
0-691-08221-9/78/0065$00.75/1 (cloth)
0-691-08213-8/78/0065$00.75/1 (paperback)

Our intention is to examine hypoellipticity of classical pseudo-differential operators with real principal symbols. Let the symbol of P be

$$p_m + p_{m-1} + \cdots$$

where p_j is positively homogeneous of degree j and p_m is real. If (x_0, ξ_0) is a simple characteristic in the sense that $p_m = 0$, $\partial p_m/\partial\xi \neq 0$ at (x_0, ξ_0) then P is not hypoelliptic. In fact, one can then construct distributions u with $Pu \in C^\infty$ near x_0 so that sing supp u is the regular bicharacteristic curve through x_0. As in section 6 of the introductory lectures by L. Hörmander the proof can be reduced to the case $P = D_1$ when it is trivial.

From now on we shall assume that

$$p_m(x,\xi) \geqq 0 \quad \text{for} \quad x \in \Omega, \ \xi \in \mathbf{R}^n .$$

This implies that there are no simple characteristics. We want to examine when P is hypoelliptic with loss of one derivative. To obtain a necessary condition we apply (1.1) for a characteristic (x_0, ξ_0) with $|\xi_0| = 1$ to

$$u(x) = v(t(x-x_0))\, e^{it^2 < x, \xi_0 >} \tag{1.2}$$

and let $t \to \infty$. Then we have

$$P(x, D)u = f(t(x-x_0))\, e^{it^2 < x, \xi_0 >}, \quad f(x) = P(x_0+x/t, t^2\xi_0+tD)\, v(x) .$$

By Taylor's formula

$$P(x_0+x/t,\ t^2\xi_0+\xi t) = t^{2(m-1)}(Q(x,\xi) + P_{m-1}(x_0, \xi_0) + 0(t^{-2}))$$

where $Q(x,\xi)$ is the quadratic form which begins the Taylor expansion of $P_m(x_0+x, \xi_0+\xi)$. Hence (1.1) implies if $\mu = 1$

$$\|v\| \leqq C\,\|(Q(x,D) + P_{m-1}(x_0,\xi_0))v\|, \quad v \in C_0^\infty(\mathbf{R}^n) . \tag{1.1$'$}$$

It is also clear that (1.1) cannot be valid for any $\mu < 1$.

It is convenient to let x and D play symmetric roles by introducing the operator $Q^s(x,D)$ where a term $2cx_jD_j$ in $Q(x,D)$ is replaced by $c(x_jD_j+D_jx_j)$. This changes the constant term in $(1.1)'$ so we define:

DEFINITION 1.2. The subprincipal symbol of P is

$$S_P(x,\xi) = p_{m-1}(x,\xi) - \frac{1}{2i}\sum_{j=1}^{n} \partial^2 p_m/\partial x_j \partial\xi_j \, . \tag{1.3}$$

A simple calculation gives

LEMMA 1.3. *The subprincipal symbol of* P *is invariantly defined at any point where* $dp_m(x,\xi) = 0$, *and it is real if* P *is formally self-adjoint.*

We shall now study when the estimate

$$\|u\| \leqq C\,\|(Q^s(x,D)+c)u\|, \quad u \in C_0^\infty(\mathbf{R}^n), \tag{1.4}$$

is valid for a non-negative quadratic form Q. To do so we shall use the invariance of (1.4) under linear symplectic transformations, that is, linear transformations preserving the symplectic form on $V = \mathbf{R}^n \times \mathbf{R}^n$ defined by

$$\sigma((x,\xi),(y,\eta)) = \langle y,\xi\rangle - \langle x,\eta\rangle \, . \tag{1.5}$$

LEMMA 1.4. *The estimate* (1.4) *remains valid with the same constant if the symbol* Q *is composed with a linear symplectic transformation.*

Proof. For every linear symplectic transformation χ there is a unitary map U such that $U^{-1}L(x,D)U = (L\circ\chi)(x,D)$ for every linear L, hence $U^{-1}Q^s(x,D)U = (Q\circ\chi)(x,D)$ for every quadratic form Q. (see e.g. Melin [9].)

The last lemma motivates us to study the extent to which the non-negative quadratic form Q can be simplified by canonical transformations. Let $Q(t,s)$ be the symmetric bilinear form on V associated to Q. Since we are going to make symplectic changes of variables, it is more natural to express this bilinear form as

$$Q(t,s) = 2\sigma(t, Fs), t, s \in V , \tag{1.6}$$

where F is skew symmetric with respect to σ. Of course F is an invariantly defined linear map on V. It is called the fundamental matrix (or Hamilton map) of Q. Let us extend F and the bilinear forms Q and σ to the complexification V_c of V.

LEMMA 1.5. $Q(t,\bar{t}) = 0 \Longleftrightarrow F(\text{Re } t) = F(\text{Im } t) = 0 \Longleftrightarrow F(t) = 0$.

Proof. If $t = u + iv$, u, v real, and

$$0 = Q(t,\bar{t}) = Q(u,u) + Q(v,v)$$

we must have $Q(u,u) = Q(v,v) = 0$ since Q is non-negative. By the Cauchy-Schwarz inequality this implies $Q(u,s) = 0$ for every s, hence $Fu = 0$. The rest is obvious.

If $\lambda \neq 0$ is an eigenvalue of F and t is a corresponding eigenvector then

$$0 < Q(t,\bar{t}) = \sigma(t, F\bar{t}) = \bar{\lambda}\sigma(t,\bar{t})$$

so λ is purely imaginary and $\text{Re } t$, $\text{Im } t$ span a symplectic two dimensional plane invariant under F. Hence the plane symplectically orthogonal to $\text{Re } t$, $\text{Im } t$ is invariant under F. Repeated use of this observation gives a symplectic basis such that

$$Q(x,\xi) = \sum_1^d \frac{1}{2}\lambda_j(x_j^2 + \xi_j^2) + Q_1(x,\xi)$$

where Q_1 is independent of $x_1, \cdots, x_d, \xi_1, \cdots, \xi_d$ and the corresponding fundamental matrix has no non-zero eigenvalue. To simplify the notations we assume for a moment that $Q = Q_1$. We have $F^2 = 0$ then for if $F^3 t = 0$ then

$$0 = \sigma(t, F^3 t) = -Q(Ft, Ft)$$

which implies $Ft \in \operatorname{Ker} F$ by Lemma 1.1. But the range of F is the orthogonal space of the kernel so $F^2 = 0$ means that $\operatorname{Ker} F$ contains its orthogonal space. Thus $\operatorname{Ker} F$ is involutive and we can choose symplectic coordinates so that $\operatorname{Ker} F$ contains the plane $x = 0$. This means that Q is a polynomial in x only, and summing up we have proved

LEMMA 1.6. *If Q is a positive semi-definite quadratic form on a real symplectic vector space then we can choose symplectic coordinates so that*

$$Q(x, \xi) = \sum_1^d \frac{1}{2} \lambda_j (x_j^2 + \xi_j^2) + \sum_{d+1}^r x_j^2$$

where $\pm i\lambda_j, \lambda_j > 0$, are the non-zero eigenvalues of the fundamental matrix of Q.

The eigenvalues of the harmonic oscillator

$$D_{x_1}^{\ 2} + x_1^{\ 2}$$

on $L^2(\mathbf{R})$ are $2\alpha + 1, \alpha \in \mathbf{N} = \{0, 1, 2, \cdots\}$. Two important corollaries follow immediately:

PROPOSITION 1.7. *If Q is a non-negative quadratic form then*

$$\frac{1}{2} \operatorname{tr}^+ F \|u\|^2 \leqq (A(x, D)u, u), \quad u \in C_0^\infty(\mathbf{R}^n), \tag{1.7}$$

where $\operatorname{Tr}^+ F = \sum_1^d \lambda_j$, λ_j being the positive eigenvalues of F/i if F is the fundamental matrix of Q. The constant in (1.7) is optimal.

PROPOSITION 1.8. *The estimate (1.4) holds if and only if*

$$c + Q(v, \bar{v}) + \sum (\alpha_j + \tfrac{1}{2})\lambda_j \neq 0 \tag{1.8}$$

for all $\alpha \in \mathbf{N}^d$, $v \in V_0$, *where* λ_j *are the positive eigenvalues of* F/i *and* V_0 *the space of generalized eigenvectors of* F *belonging to the eigenvalue* 0. *Here* F *is the fundamental matrix.*

The statement of the result was made unnecessarily complicated in order that it remain valid when Q is not non-negative but only takes its values in a proper cone $\Gamma \subset \mathbf{C}\setminus 0$.

From Proposition 1.7 one can return to the pseudo-differential operator P by means of a partition of unity decomposing an arbitrary function into a sum of functions which roughly speaking are of the form (1.2). This gives

THEOREM 1.9 (Melin). *Suppose that* $p_m(x,\xi) \geqq 0$ *in* $\Omega \times \dot{\mathbf{R}}^n$ *and that*

$$\mathrm{tr}^+ p_m''/2 + \mathrm{Re}\, S_P(x,\xi) \geqq 0 \quad when \quad p_m(x,\xi) = 0 .$$

For any $\varepsilon > 0$ *and* $K \subset\subset \Omega$ *there is then a constant* C *such that*

$$\text{(1.9)} \qquad \mathrm{Re}\,(P(x,D)u,u) + \varepsilon \|u\|_{(m-1)/2}^2 \geqq -C\|u\|_{m/2-1}^2, \quad u \in C_0^\infty(K) .$$

After proving a more delicate version of the extended version of Proposition 1.8 indicated above, one can also prove

THEOREM 1.10 (Boutet de Monvel et al. [4], Hörmander [7]). *Suppose that* p_m *takes its values in a proper cone* $\Gamma \subset \mathbf{C}\setminus 0$, *that the zeros form a smooth manifold* Σ *and that* $|p_m|$ *is locally bounded from below by the square of the distance to* Σ. *Then* P *is hypoelliptic with loss of one derivative if and only if for every* $\rho \in \Sigma$

$$\text{(1.10)} \qquad S_p(\rho) + Q_\rho(v,\bar v) + \sum(\alpha_j + \tfrac{1}{2})\lambda_j \neq 0, \quad v \in V_0,\ \alpha \in \mathbf{N}^{d_\rho},$$

where Q *is the quadratic form starting the Taylor expansion of* p_m *at* ρ, V_0 *the space of generalized eigenvectors of its fundamental matrix* F *belonging to the eigenvalue* 0 *and* λ_j *the* d_ρ *eigenvalues of* F/i *in* Γ.

EXAMPLES. i) $P = |D_{x''}|^2 + |x''|^2|D_{x'}|^2 + c(|D_{x'}|^2 + |D_{x''}|^2)^{1/2}$, $x' \in \mathbf{R}^{n'}$, $x'' \in \mathbf{R}^{n''}$, the Grušin operator, is hypoelliptic if and only if $c + n'' + 2\alpha \neq 0$ when $\alpha \in \mathbf{N}$.

ii) $P = |D_x|^2 + c(|D_x|^2 + |D_t|^2)^{1/2}$, $x \in \mathbf{R}^n, t \in \mathbf{R}$, is hypoelliptic if and only if c is not real $\leqq 0$.

The proof of Theorem 1.13 by estimates, alluded to above, is due to Hörmander [7], who also characterized hypoellipticity with loss of one derivative in general assuming only the above cone condition. Boutet de Monvel, Grigis and Helffer [4] proved Theorem 1.13 by constructing a parametrix. This followed earlier work of Boutet de Monvel-Trèves [3], Boutet de Monvel [2], Sjöstrand [12] and Grigis.

2. *The asymptotic behavior of the eigenvalues*

Let X be a compact n-dimensional C^∞ manifold without boundary, let dx be a fixed positive density on X and let P be a classical pseudo-differential operator on X of order $m > 0$ with symbol $\sim p(x,\xi) + p_{m-1}(x,\xi) + \cdots$ where $p_j(x,\xi)$ is positively homogeneous of degree j in ξ and where $p = p_m$ is the principal symbol. Suppose that $p(x,\xi) \geq 0$ and that P is formally self-adjoint, that is

$$\int Pu\,\bar{v}\,dx = \int u\,\overline{Pv}\,dx, \qquad u, v \in C^\infty(X).$$

When P is elliptic the spectrum of P is discrete and bounded from below. Denoting by $N(\lambda)$ the number of eigenvalues $\leq \lambda$, it is well known that

$$N(\lambda) \sim (2\pi)^{-n} \int_{p(x,\xi) \leq 1} dx d\xi \, \lambda^{n/m} \quad \text{as} \quad \lambda \to \infty \tag{2.1}$$

(see [6]).

In this lecture we shall describe some joint work with J. Sjöstrand on the distribution of eigenvalues for a class of the hypoelliptic operators discussed above. We shall suppose

1°) $m > 1$ and $p(x,\xi) \geq 0$

2°) $\Sigma = p^{-1}(0) \subset T^*X \setminus 0$ is a symplectic submanifold of codimension $2d$ and p vanishes to precisely second order on Σ, that is, for all $K \subset\subset T^*X \setminus 0$ there is a constant such that $p(x,\xi) \geq C_K d(x,\xi)^2$, where d denotes the distance of (x,ξ) to Σ.

3°) $S_P(\rho) + \frac{1}{2}\,\mathrm{tr}^+ F_\rho > 0$ for $\rho \in \Sigma$, where S_P is the subprincipal symbol of P and F_ρ is the fundamental matrix of p, that is of the Hessian quadratic form $p''_\rho(t,t)$ on $T_\rho(T^*X)$ of the second order terms in the Taylor expansion.

That Σ is symplectic means that the standard symplectic form $\sigma = \Sigma\, d\xi_j \wedge dx_j$ restricted to $T(\Sigma)$ is non-degenerate. As a consequence the quadratic form $p''_\rho(t,t)$ restricted to $N(\Sigma) = T(\Sigma)^\perp$ is positive definite. Since it is zero on $T(\Sigma)$, its fundamental matrix F_ρ has no nilpotent part. Condition 3° implies P is hypoelliptic with loss of one derivative. From this we may conclude that P is self-adjoint (as an unbounded operator) with domain $\mathfrak{D} = \{u \in L^2(X) : Pu \in L^2(X)\}$ and since $m > 1$ that the spectrum of P is discrete. To see the latter, let E be any orthogonal projection on a bounded part of the spectrum of P. Since $P^k Eu \in L^2$ for $u \in L^2$ it follows from hypoellipticity that $E : L^2 \to C^\infty$. Applying the estimate

$$\|u\|_{m-1} \leq C(\|Pu\|_0 + \|u\|_0), \qquad u \in C^\infty(X)$$

to Eu we get

$$\|Eu\|_{m-1} \leq C(\|PEu\|_0 + \|Eu\|_0) \leq C'\|Eu\|_0, \qquad u \in C^\infty(X)\,.$$

From the last inequality and the local compactness of H_{m-1} in H_0 we deduce that E has finite rank and consequently that P has discrete spectrum.

Condition 3° is actually stronger than hypoellipticity. Using Melin's theorem we may find two self-adjoint elliptic operators A and B bounded from below and of order $m-1$ and m respectively such that $A < P < B$.

This tells us that P is bounded from below. Furthermore, defining $N(\lambda)$ as before we get for some constant $C > 0$

(2.2) $$C^{-1}\lambda^{n/m} \leq N(\lambda) \leq C\lambda^{n/(m-1)}, \quad \lambda > C .$$

This follows from (2.1) and the max-min principle.

We have a more precise result, namely

THEOREM 2.1. *Assume* 1°, 2°, and 3°. *Then*

(i) *if* $md-n>0$

$$N(\lambda) \sim \frac{1}{(2\pi)^n} \int_{p(x,\xi)\leq 1} dxd\xi \, \lambda^{n/m} \quad \text{as} \quad \lambda \to \infty$$

(ii) *if* $md-n=0$

$$N(\lambda) \sim \frac{1}{(2\pi)^{n-d}\, m\,(d-1)!} \iint_{\Sigma\cap\{\det p''_{(x'',\xi'')}\leq 1\}} dx'd\xi' \, \lambda^{n/m} \log\lambda \quad \text{as} \quad \lambda \to \infty$$

(iii) *if* $md-n<0$ *and* $S_{P,\alpha} = S_P + \Sigma\,(2\alpha_j+1)\lambda_j/2$

$$N(\lambda) \sim \frac{1}{(2\pi)^{n-d}} \left(\sum_{\alpha\in N^d} \iint_{\Sigma\cap\{S_{P,\alpha}\leq 1\}} dx'd\xi' \right) \lambda^{\frac{n-d}{m-1}}, \quad \lambda \to \infty .$$

Here $\pm i\lambda_j$ *are the non-vanishing eigenvalues of* F_ρ, $\lambda_j > 0$ *and* (x',ξ') *are symplectic coordinates on* Σ, (x'',ξ'') *symplectic coordinates in* $N(\Sigma)$.

We remark that if (x,ξ) are symplectic coordinates on T^*X, then $dxd\xi = \pm(1/n!)\sigma^n$ is an invariantly defined density. Similarly, $dx'd\xi'$ is invariantly defined on Σ, as well as all the other expressions.

A second remark is that the integral in (i) converges precisely when $md-n>0$. The asymptotic formula for $N(\lambda)$ is the same as in the elliptic case as long as it makes sense. The sum in (iii) converges precisely when $md - n < 0$. Case (ii) may be thought of as a transitionary case since when $md - n = 0$ we have $(n-d)/(m-1) = n/m$.

Since $\operatorname{tr} e^{-tP} = \int e^{-\lambda t} dN(\lambda)$, Theorem 2.1 will be a consequence of Karamata's Tauberian theorem if we can find the singularity of $\operatorname{tr} e^{-tP}$ as $t \downarrow 0$. To do so we'll approximate e^{-tP} by a Fourier integral operator with a complex phase function. We want to solve

$$(2.3) \qquad \begin{cases} D_t v = iP(x, D)v & \text{in} \quad \mathbf{R}^+ \times X \\ v|_{t=0} = u & u \in C_0^\infty(X) . \end{cases}$$

We'll discuss how to do this locally for a coordinate patch $\Omega \subset X$. We look for a solution of the form

$$(2.4) \qquad v(x,t) = A_t u(x) = (2\pi)^{-n} \int e^{i\phi(t,x,\eta)} a(t,x,\eta)\hat{u}(\eta)\, d\eta .$$

We'll say that a function $b(t,x,\eta)$ is quasi-homogenous (q.h.) of degree k if $\lambda^k b(\lambda^{m-1}t, x, \eta) = b(t, x, \lambda\eta)$. We'll want ϕ to be q.h. of degree 1 and $a \sim a_0 + a_1 + \cdots$ where a_j is q.h. of degree $-j$. (This essentially amounts to making the change of time scale $s = |\eta|^{m-1}t$.) Applying $D_t - iP$ to A_t, using formally the standard asymptotic expansion and grouping the terms by q.h., we get the characteristic equation

$$(2.5) \qquad \phi_t' = ip(x, \phi_x'), \quad \phi(0,x,\eta) = \langle x, \eta \rangle$$

and a sequence of transport equations. We will of course require that $\operatorname{Im} \phi \geq 0$. To solve (2.3) modulo a smoothing operator we need only solve (2.5) modulo $O((\operatorname{Im}\phi)^N)$ for all N, locally uniformly.

We'll limit ourselves to a discussion of only the characteristic equation. Just to see what's going on, we solve (2.5) in the easy case that P is elliptic. Making the substitution $s = t|\eta|^{m-1}$ converts (2.5) into the homogenous equation

(2.6) $$\phi'_s = ip(x, \phi'_x)/|\eta|^{m-1}, \quad \phi(0, x, \eta) = <x, \eta> .$$

Trying a power series in s we find

$$\phi(s, x, \eta) = <x, \eta> + isp(x, \eta)/|\eta|^{m-1} + s^2\psi_2(x, \eta) + \cdots$$

which solves (2.6) modulo $O(|\eta|s^N)$ for any N. For s small $\operatorname{Im} \phi \geq O(|\eta|s)$. By modifying ψ_2, etc., for large s we can get $\operatorname{Im} \phi \geq O(|\eta|s)$ for all s and our solution is then good modulo $O((\operatorname{Im} \phi/|\eta|)^N)$ for all N.

Returning to the non-elliptic situation, we may find a solution of

(2.7) $$\phi'_t - ip(x, \phi'_x) = O(|\operatorname{Im} \phi|^N), \quad \phi(0, x, \eta) = <x, \eta>$$

for any finite interval $0 \leq t \leq T$ using the methods of Trèves [13] or Melin-Sjöstrand [11]. Furthermore, we will have

(2.8) $$\operatorname{Im} \phi(t, x, \eta) \geq ctd(x, \eta)^2$$

where $d(x, \xi)$ is the distance to Σ. The crucial step in our analysis is to show

PROPOSITION 2.2. *There exists* $\phi(\infty, x, \eta) \in C^\infty(\Omega \times \mathbf{R}^n)$ *such that*

(i) $\operatorname{Im} \phi(\infty, x, \eta) \geq C(d(x, \eta))^2$

(ii) *for every* $\lambda \leq \min \lambda_j$, *for all multi-indices* α, β, γ, *and* $K \subset\subset \Omega \times \mathbf{R}^n$ *there is a constant* C *such that on* $\overline{\mathbf{R}}_+ \times K$

$$|\partial_x^\alpha \partial_\eta^\beta \partial_t^\gamma (\phi(t, x, \eta) - \phi(\infty, x, \eta))| \leq Ce^{-\lambda t} .$$

Because of (i) it will be sufficient for us to argue using Taylor expansions in the directions transversal to Σ. We shall reason geometrically. Let $H_{p(x,\xi)} = \Sigma(\partial p/\partial \xi_j)\partial/\partial x_j - (\partial p/\partial x_j)\partial \xi_j$ be the Hamiltonian vector field on $\widetilde{T^*X} \times \widetilde{T^*Y}$ where Y is a second copy of X and H_p only acts on the first factor. We shall consider H_p to be the real vector field on the complexification of T^*X which coincides with H_p when applied to almost analytic functions. Let C_t be the graph of the canonical relation $\exp(-itH_p)$

$$C_t = \{(\exp(-itH_p)(\rho), \rho);\ \rho \in T^*X\setminus 0\}\ . \tag{2.9}$$

If $\phi(t,x,\eta)$ is a generating function for the canonical relation C_t, that is, $C_t = \{(x, \phi'_x, \phi'_\eta, \eta)\}$ then $\phi(t,x,\eta)$ will be the desired solution of (2.7).

It is not hard to find local (complex) symplectic coordinates (x,ξ) on $\widetilde{T^*X}$ so that Σ is given by $\tilde{x}'' = \tilde{\xi}'' = 0$ and

$$p = i\langle A(\tilde{x},\tilde{\xi})\tilde{x}'', \tilde{\xi}''\rangle + O(|(\tilde{x}'',\tilde{\xi}'')|^3) \overset{\text{def}}{=} \tilde{p}(\tilde{x},\tilde{\xi}) \tag{2.10}$$

where the matrix A has only positive eigenvalues for $\tilde{x}'' = \tilde{\xi}'' = 0$. In these coordinates the fundamental matrix is

$$F = i\begin{pmatrix} A & 0 \\ 0 & -{}^tA \end{pmatrix}$$

so the eigenvalues of A are $\lambda_1, \cdots, \lambda_d$. Considering Taylor expansions in $\tilde{x}''$ one can construct $\Phi = \Phi(\tilde{x}, \tilde{\eta}')$ so that

$$\begin{aligned} \tilde{p}(\tilde{x}, \Phi'_{\tilde{x}}) &= O(|\tilde{x}''|^N), \quad \forall N \\ (\tilde{x}', 0, \tilde{\eta}') &= \langle \tilde{x}', \tilde{\eta}'\rangle\ . \end{aligned} \tag{2.11}$$

If we define (formally and locally)

$$J_+ = \{(\tilde{x}, \Phi'_x(\tilde{x},\tilde{\eta}'));\ \tilde{x} \in \mathbf{C}^n,\ \tilde{\eta}' \in \mathbf{C}^{n-d}\}\ ,$$

then in the original coordinates J_+ is a well defined almost analytic involutive manifold in $\widetilde{T^*\Omega}\setminus 0$ with $J_+ \cap T^*\Omega\setminus 0 = \Sigma$. Put $J_- = \overline{J_+}$.

We can next modify slightly the coordinates $(\tilde{x},\tilde{\xi})$ so that J_+ is given by $\tilde{\xi}'' = 0$, and J_- by $\tilde{x}'' = 0$. Then (2.10) improves to (with a new matrix A for $(\tilde{x}'', \tilde{\xi}'') \neq 0)$:

$$(2.10)' \qquad \tilde{p}(\tilde{x}, \tilde{\xi}) = i\langle A(\tilde{x}, \tilde{\xi})\tilde{x}'', \tilde{\xi}''\rangle + O(|(\tilde{x}'', \tilde{\xi}'')|^N), \ \forall N .$$

We can now study

$$\partial\tilde{\phi}/\partial t - i\tilde{p}(\tilde{x}, \tilde{\phi}'_{\tilde{x}}) = O(|(\tilde{x}'', \tilde{\eta}'')|^N), \ \forall N$$

$$\tilde{\phi}|_{t=0} = \langle \tilde{x}, \tilde{\eta}\rangle$$

by Taylor expansion in (x'', η''). The solution is of the form

$$\tilde{\phi}(t, \tilde{x}, \tilde{\eta}) = \langle \tilde{x}', \tilde{\eta}'\rangle + \langle e^{-tA}\tilde{x}'', \tilde{\eta}''\rangle + O(|(\tilde{x}'', \tilde{\eta}'')|^3) ,$$

$A = A(\tilde{x}', 0, \tilde{\eta}', 0)$ and $\tilde{\phi}$ turns out to converge to $\langle \tilde{x}', \tilde{\eta}'\rangle$ exponentially fast in the sense of Proposition 2.2. The limiting canonical relation is $C_\infty = \{(\tilde{x}, \tilde{\eta}', 0, \tilde{x}', 0, \tilde{\eta})$ and (in the original coordinates) it can easily be verified that C_∞ is strictly positive in the sense of [9] with real part $\Delta(\Sigma\times\Sigma)$. Proposition 2.2 then follows rather easily.

The transport equations can be treated similarly and the assumption 3° implies that all the quasi-homogeneous terms in the expansion of the amplitude are exponentially decreasing with respect to t. The operator (2.4) can therefore be given a meaning and will approximate e^{-tP}.

After having solved the characteristic equation and the transport equations we need to compute

$$(2.12) \qquad \operatorname{tr}(e^{-tP}) = (2\pi)^{-n} \iint_{T^*X} e^{i(\phi(t,x,\eta) - \langle x,\eta\rangle)}\, a(t, x, \eta)\, dx d\eta .$$

In the most interesting case where $md - n < 0$ the following computations can be justified. We write $a \sim a_0 + a_1 + \cdots$ and replace η by $t^{-1/(m-1)}\eta$. Then by quasi-homogeneity

$$(2.13) \quad \operatorname{tr}(e^{-tP}) \sim (2\pi)^{-n} t^{-n/(m-1)} \sum \iint e^{i\psi(t,x,\eta)}\, t^{j/(m-1)} a_j(1,x,\eta)\, dx d\eta$$

where $\psi(t,x,\eta) = t^{-1/(m-1)}(\phi(1,x,\eta) - \langle x,\eta\rangle)$. The critical points of the non-degenerate phase function ψ are on Σ. This suggests choosing symplectic coordinates θ such that Σ is given by $\theta'' = 0$ and evaluating the integral with respect to θ'' by the method of stationary phase, considering $t^{-1/(m-1)}$ as the large parameter. The leading term is

$$(2.14) \qquad (2\pi)^{d-n} t^{-\left(\frac{n-d}{m-1}\right)} \iint_{\Sigma} \frac{a_0(1,\theta',0)\, d\theta'}{(\det i \operatorname{Hess}_{\theta''}(\phi(1,x,\eta) - \langle x,\eta\rangle))^{1/2}} .$$

Evaluating this integral gives the result of case (iii) in Theorem 2.1.

EXAMPLE. For the Grušin operator $P = D_{x''}^2 + x''^2 |D_{x'}|^2 + c|D_{x'}|$, $x \in \mathbf{R}^n$, $x' \in \mathbf{R}^d$, we have $\Sigma = \{x'' = \xi'' = 0\}$. Near Σ the exact solution of the characteristic equation

$$\begin{cases} \partial\phi/\partial t = i[|\phi'_{x''}|^2 + |x''|^2 |\phi'_{x'}|^2] \\ \phi(0,x,\eta) = \langle x,\eta\rangle \end{cases}$$

is

$$\phi(t,x,\eta) = \langle x',\eta'\rangle + \frac{\langle x'',\eta\rangle}{\cosh(2t|\eta'|)} + i\frac{|\eta'|}{2}\tanh(2t|\eta'|)(|x''|^2 + |\eta''/|\eta'||^2).$$

Looking for an amplitude independent of x, we get the transport equation

$$D_t a(t,\eta) = i \quad c|\eta'| + \frac{1}{i}\sum \frac{\partial^2\phi}{\partial x_j''^2} \quad a(t,\eta)$$

whose exact solution on Σ is

$$a(t,\eta) = e^{-ct|\eta'|}(\cosh(2t|\eta'|))^{-d/2} .$$

For certain second order operators Metivier [11] has obtained formulas for the spectral function. Gaveau [5] has studied the heat kernel of the Kohn-Laplacian on Heisenberg groups. Bolley, Camus and Pham [1] have obtained the distributions of eigenvalues of certain elliptic operators degenerating on the boundary of a domain. Trèves has also constructed the heat kernel for operators with double characteristics by a different method.

REFERENCES

[1] C. Bolley, J. Camus, and T. Pham, Journées de EDP à St. Jean des Monts (1977), Springer LN 660, 33-46.

[2] L. Boutet de Monvel, Hypoelliptic operators with double characteristics and related pseudo-differential operators, Comm. Pure Appl. Math. 27 (1974), 585-639.

[3] L. Boutet de Monvel and F. Trèves, On a class of pseudo-differential operators with double characteristics, Invent. Math. 24 (1974), 1-34.

[4] L. Boutet de Monvel, A. Grigis, B. Helffer, Paramétrixes d'opérateurs pseudo-différentiels a caractéristique multiples Astérisque,

[5] B. Gaveau, Principe de moindre action, propagation de la chaleur et estimées sous elliptiques sur certains groups nilpotent, Acta Math. 139 (1977),

[6] L. Hörmander, The spectral function of an elliptic operator, Acta Math. 121 (1968), 193-218.

[7] ————, A class of hypoelliptic operators with double characteristics, Math. Ann. 217 (1975), 165-188.

[8] A. Melin, Lower bounds for pseudo-differential operators, Ark. för Mat. 9 (1971), 117-140.

[9] A. Melin and J. Sjöstrand, Fourier integral operators with complex valued phase function, Springer LN 459 (1975), 120-233.

[10] ————, Fourier integral operators with complex phase functions and a parametrix for an interior boundary value problem, Comm. P.D.E. 1 (1976), 313-400.

[11] G. Metivier, Fonction spectrale et valeurs propres d'une class d'operateurs non-elliptiques, Comm. P.D.E. 1 (1976), 465-514.

[12] J. Sjöstrand, Parametrices for pseudo-differential operators with multiple characteristics, Arkiv. för Mat. 12 (1974), 85-130.

[13] F. Trèves, Solution of Cauchy problems modulo flat functions, Comm. P.D.E. 1 (1976), 45-72.

DIFFERENTIAL BOUNDARY VALUE PROBLEMS OF PRINCIPAL TYPE

R. B. Melrose[1)]

1. *Introduction*

In these lectures our main concern is with the behavior, in particular the singularities, of solutions to boundary value problems for linear partial differential operators. Thus, we are given a C^∞ manifold M (Hausdorff, paracompact) with boundary ∂M. On M is prescribed a linear partial differential operator P. We shall make the following assumptions on P:

(1.1) P is second order and of real principal type.

(1.2) ∂M is nowhere characteristic for P.

Both these conditions are restrictions only on the principal symbol, p, of P. Thus, (1.1) requires that $p \in C^\infty(T^*M)$ be real-valued and that its zero surface be smooth and non-radial

(1.3) dp, α linearly independent on $p = 0$ in $T^*M\backslash 0$,

where α is the fundamental 1-form on T^*M. In fact, within the omnibus term 'real principal type' we shall include a stronger version of (1.3) at the boundary:

1)Supported in part by NSF grant MCS77-18723.

Seminar on Singularities of Solutions
0-691-08221-9/78/0081 $01.55/1 (cloth)
0-691-08213-8/78/0081 $01.55/1 (paperback)

(1.4) dp, α restricted to $\partial T^*M\backslash 0$ are independent on $p = 0$.

The significance of these restrictions will, I hope, become clearer later, but (1.2) is fairly immediate; it means simply

$$p \neq 0 \quad \text{on} \quad N^*\partial M\backslash 0 \tag{1.5}$$

where $N^*\partial M$ is the conormal bundle to ∂M. By local coordinates at $m \in \partial M$ we shall always mean coordinates $(x, y) \in \mathbf{R}\times\mathbf{R}^n$ such that $x \geq 0$ exactly on M and $x = 0$, $y = 0$ at m. The corresponding (dual) coordinates in T^*M will be denoted (x, y, ξ, η). Now, p is a real polynomial homogeneous of degree 2 and $N^*\partial M = \{(0, y, \xi, 0)\}$ so (1.5) states that

$$p = g(x, y)(\xi^2 + \xi a(x, y, \eta) + b(x, y, \eta))$$

with $g \neq 0$ at $x = 0$.

(1.6) LEMMA. *If* ∂M *is non-characteristic at* $m \in \partial M$ *for* P *of real principal type, then there exist local coordinates* (x, y) *at* m *such that*

$$p = \pm(\xi^2 + r(x, y, \eta)) .$$

Moreover, these coordinates are unique up to an x*-independent change in the* y*-coordinates.*

Now, we are interested in solutions, u, to

$$Pu = f \in C^\infty(M) \tag{1.7}$$

where $C^\infty(M)$ is the space of functions with all derivatives continuous up to the boundary, $u \in \mathcal{D}'(M)$ is an extendible distribution. (Throughout these lectures we shall assume, for simplicity, that M carries a smooth, non-vanishing *density* which will be used to trivialize all density bundles. Thus $\mathcal{D}'(M)$ is the dual to $C_0^\infty(M)$, the space of compactly supported C^∞ functions which vanish, with all derivatives on ∂M.) Such an equation,

(1.7), does not impose enough of a restriction on u near ∂M so we also impose a *boundary condition.*

Let B be a differential operator on M, near ∂M, of order 0 or 1 with

(1.8) $$\partial M \text{ non-characteristic for } B .$$

Then, in addition to (1.7) we shall suppose

(1.9) $$Bu|_{\partial M} = v \in C^\infty(\partial M)$$

or possibly some weaker condition on the singularities of v. As a consequence of (1.2), (1.7), (1.9) does indeed make sense.

(1.10) THEOREM (Peetre, Hörmander [7]). *If* ∂M *is non-characteristic for* P *(of any order) and* $Pu \in C^\infty(M)$ *then in any local coordinate system at* $m \in \partial M$,

$$u \in C^\infty(\overline{\mathbf{R}^+_x}; \mathcal{D}'(\mathbf{R}^n_y)) .$$

For the main part we shall consider the Dirichlet boundary condition B = 1. For other choices of B one must require certain conditions relating P and B.

2. *Examples*

The following three classical second-order boundary value problems of principal type exemplify the theorems and constructions to be carried out. The first two are simpler than the third, which to a considerable extent guides our analysis.

(2.1) M = X, a compact Riemann manifold with boundary, $P = \Delta$, the Laplace Beltrani operator (B = 1), e.g. solve

$$\Delta u = 0$$
$$u|_{\partial M} = v .$$

Classically one knows existence, uniqueness and regularity results for this problem.

(2.2) $M = \overline{\mathbf{R}_t^+} \times Y$, Y a compact Riemann manifold without boundary

$P = \partial_t^2 - \Delta$, the wave operator $(B = 1)$

$$(\partial_t^2 - \Delta)u = 0$$
$$u|_{t=0} = v \in \mathcal{D}'(Y)$$
$$(\partial_t u|_{t=0} = w \in \mathcal{D}'(Y)) .$$

By adding the extra boundary condition, this becomes the Cauchy problem; existence, uniqueness and propagation of singularities results then follow by standard (classical, pseudo-differential or Fourier integral operator) techniques (see Hörmander [8]). The problem without the extra boundary condition is then easily analyzed.

(2.3) $M = \mathbf{R}_t \times X$, X a Riemann manifold with boundary. We shall also require that X be either compact or else be Euclidean outside a compact set $(B = 1)$

$$(\partial_t^2 - \Delta)u = 0$$
$$u|_{\partial M} = v$$
$$(u|_{t=0} = u_0,\ \partial_t u|_{t=0} = u_1) .$$

With the extra (initial) conditions added there is existence and uniqueness and regularity of solutions (given compatibility conditions). The main question of interest concerns the singularities of u.

3. *Symplectic geometry*

As has been emphasized, for both the C^∞ and analytic categories, in the preceding lectures one has in the analysis of differential equations the freedom to make canonical transformations in $T^*M \backslash 0$, to simplify the problem. In the past this approach has not been used systematically in the study of boundary value problems and it is just such a treatment which will be considered here.

If $\rho \in T^*\dot{M}\backslash 0$ $(\dot{M} = M\backslash\partial M)$ and $p(\rho) \neq 0$, i.e. ρ is an elliptic point, the equation $Pu = f$ can be microlocally inverted (see Hormander [9]): $u = Qf + g$ where $\rho \notin SS(g)$ and $Q \in L^{-2}(\dot{M})$. If $p(\rho) = 0$, $\rho \in T^*\dot{M}\backslash 0$, then as a consequence of our assumption that P is of real principal type, there is a Fourier integral operator F, associated to a canonical transformation

$$\chi : T^*\dot{M}\backslash 0, \rho \to T^*\mathbf{R}^{n+1}\backslash 0, (0, \tilde{\theta}) ,$$

$\theta = (0, \cdots, 0, 1)$, such that

$$\partial_{z_1}(Fu) = FQf + g ,$$

$\rho \notin SS(g)$, $Q \in L^{-1}(\dot{M})$. Combining these results gives

(3.1) THEOREM (Hörmander, see [9]). *If* $Pu \in C^\infty(\dot{M})$ *then* $SS(u|_{\dot{M}}) \subset \Sigma(P) = \{\rho \in T^*\dot{M}\backslash 0;\ p(\rho) = 0\}$ *is invariant under the Hamilton flow,* $\exp(tH_p)$, *of* p.

This does not explain what happens to singularities when the bicharacteristics (integral curves of H_p in Σ) meet the boundary $\partial T^*M\backslash 0$. The first obvious question is: what is a canonical transformation on a manifold with boundary?

If $\rho \in \partial T^*M\backslash 0$ such a (germ of) transformation

$$\chi : T^*M\backslash 0, \rho \to T^*(\overline{\mathbf{R}^+_x} \times \mathbf{R}^n_y)\backslash 0, \tilde{\rho}$$

should satisfy the usual conditions: it should be homogeneous and canonical

$$\chi^*\left(\xi\, dx + \sum_{j=1}^{n} \eta_j\, dy_j\right) = \alpha_M$$

and should preserve the boundary:

$$\begin{array}{ccc} \chi : \partial T^*M\backslash 0,\ \rho & \longrightarrow & \partial T^*(\overline{\mathbf{R}^+}\times \mathbf{R})\backslash 0,\ \tilde{\rho} \\ \Big\downarrow \iota^* & & \Big\downarrow \\ \partial\chi : T^*\partial M\backslash 0,\ \sigma & \longrightarrow & T^*\mathbf{R}^n\backslash 0,\ \tilde{\sigma} \end{array} \tag{3.2}$$

Since χ is canonical it must preserve the Hamilton foliation of the boundary, and since the leaves are just the fibers of the projection $\iota^* : \partial T^*M \to T^*\partial M$ induced from $\iota : \partial M \hookrightarrow M$, χ must project to a canonical transformation $\partial\chi$ on the boundary, giving a commutative diagram (3.2).

Now, the projection ι^* is an affine line bundle over $T^*\partial M$ and we shall say that χ is a *boundary canonical transformation* if χ in (3.2) is an affine-linear bundle map and if it satisfies, in addition, all the analogous affinity conditions on the higher order jet spaces. Rather than describe these in detail, let us simply note how they appear in coordinates:

$$\chi : T^*(\overline{\mathbf{R}^+_x}\times \mathbf{R}^n_y)\backslash 0,\ \tilde{\rho} \to T^*(\overline{\mathbf{R}^+_x}\times \mathbf{R}^n_y)\backslash 0,\ \tilde{\rho}$$

is a boundary canonical transformation if it is canonical, $\chi^*(x) = 0$ on $x = 0$ and the function $\Xi(x,y,\xi,\eta) = \chi^*(\xi)$ satisfies:

$$\frac{\partial^k}{\partial x^k}\Xi(0,y,\xi,\eta) \text{ is a polynomial of degree } \le k+1 \text{ in } \xi . \tag{3.3}$$

These conditions (3.3) are easily seen to be coordinate-free since, under a coordinate transformation preserving $x = 0$, ξ is replaced by a linear function of ξ and the new η's are linear functions too, with the coefficient of ξ vanishing at $x = 0$.

Now, given $p \in C^\infty(T^*M)$, we shall define an equivalence relation on $T^*\partial M\backslash 0$ as follows: $\sigma_1 \sim \sigma_2$ if there exists a boundary canonical transformation χ such that $\partial\chi(\sigma_1) = \sigma_2$ and $\chi(\Sigma(P)) \subset \Sigma(P)$ near $\iota^{*-1}(\sigma_2)$. Since χ preserves the fibers of (3.2) the number of points in

$\Sigma(P) \cap \iota^{*-1}(\sigma)$ is an invariant of this equivalence relation. Accordingly, we consider the following classification:

$$\mathcal{E} = \{\sigma \in T^*\partial M\backslash 0;\ \Sigma(P) \cap \iota^{*-1}(\sigma) = \emptyset\}$$

$$\mathcal{H} = \{\sigma \in T^*\partial M\backslash 0;\ \Sigma(P) \cap \iota^{*-1}(\sigma) = \{\rho^+, \rho^-\},\ \rho^+ \neq \rho^-\}$$

$$\mathcal{G} = \{\sigma \in T^*\partial M\backslash 0;\ \Sigma(P) \cap \iota^{*-1}(\sigma) = \{\rho\}\}\ .$$

In the local coordinates introduced in Lemma 1.6, $\sigma = (y, \eta) \in \mathcal{E}, \mathcal{H}, \mathcal{G}$ as $r_0 = r(0, y, \eta) >, <, = 0$. Then (1.4) becomes

$$dr_0,\ a_{\partial M} \text{ lin. indep. on } r_0 = 0\ . \tag{3.4}$$

Thus $\mathcal{G} \subset T^*\partial M\backslash 0$ is a non-radial, conic, hypersurface which plays, for boundary value problems, a role similar to that played by $\Sigma(P)$ in open regions. In the examples of section 2, note that, in (2.1) $\mathcal{E} = T^*\partial M\backslash 0$ — all boundary points are elliptic and in (2.2), $\mathcal{H} = T^*\partial M\backslash 0$ — all boundary points are hyperbolic. For the final example, (2.3), $\mathcal{G} \neq \emptyset$, is a hypersurface separating $\mathcal{H}$ and $\mathcal{E}$.

A complete answer to the equivalence problem posed above has not been obtained and is not to be expected, at least in a simple form. However, the known equivalence results, which concern the 'non-degenerate' points can be combined with other methods to yield a fairly complete description of the singularities of u.

Using the observations above one can easily see that

$$\mathcal{G} = \{\sigma \in T^*\partial M\backslash 0;\ \exists \rho \in \Sigma(P) \cap \iota^{*-1}(\sigma) \text{ with } \{p, x\}(\rho) = 0\}\ .$$

Here $\{f, g\}$ is the Poisson bracket on T^*M, is equal to $H_f g$. Thus, $\{p, x\} = 0$ means that H_p, the Hamilton field of p is tangent to ∂T^*M, the boundary. Thus $\mathcal{G}$ is the set of *glancing points*, or points of bicharacteristic tangency. It is the appearance of these points which takes the analysis of boundary value problems beyond the usual techniques. Define

$$\mathcal{G}_d \text{ (resp. } \mathcal{G}_g) = \{\sigma \in \mathcal{G};\ \{p,\{p,x\}\} > \text{(resp. } <) 0\},$$

the sets of *diffractive* (resp. *gliding*) points, at which H_p is simply tangent to ∂T^*M. $\mathcal{G}_d \cup \mathcal{G}_g$ is the 'non-degenerate' part of $\mathcal{G}$.

(3.5) THEOREM. *$\mathcal{E}$, $\mathcal{H}$, $\mathcal{G}_d$ and $\mathcal{G}_g$ are all equivalence classes under the relation on $T^*\partial M \backslash 0$ introduced above.*

There is in fact nothing to prove for points in $\mathcal{E}$. For $\mathcal{H}$ the proof follows from a careful application of Darboux's theorem (cf. [9]). For $\mathcal{G}_d$ and $\mathcal{G}_g$ the proof is of a somewhat different order, the main elements are contained in [12]. Given $\sigma_1, \sigma_2 \in \mathcal{H}$, the conjugating map χ can be chosen to have a given boundary map $\partial\chi$, provided it is canonical and has $\partial\chi(\sigma_1) = \sigma_2$. For points in, say, $\mathcal{G}_g$ this is no longer true, $\partial\chi$ must be chosen to map $\mathcal{G}$ to itself and moreover to intertwine the *billiard ball maps*, defined locally near $\mathcal{G}_g$. $\delta^+ : \bar{\mathcal{H}}, \mathcal{G}_g \to \bar{\mathcal{H}}, \mathcal{G}_g$ is defined near $\sigma_0 \in \mathcal{G}_g \subset \partial\bar{\mathcal{H}}$ $(\bar{\mathcal{H}} = \mathcal{H} \cup \mathcal{G})$ as follows:

$$\begin{array}{ccccccc} & & \partial T^*M\backslash 0 & \xrightarrow{H_p \text{ flow}} & \partial T^*M\backslash 0 & & \\ & & \cup & & \cup & & \\ \mathcal{H} \ni \sigma & \longmapsto & \rho_+ & \longmapsto & \rho & \overset{\iota^*}{\longmapsto} & \delta^+(\sigma) \end{array}$$

where ρ_+ is the point in $\iota^{*-1}(\sigma) \cap \Sigma(P)$ on the positive side. δ^+ preserves $\mathcal{G}_g$, and turns out to have a simple square root singularity on $\mathcal{G}_g$, and to be C^∞ in $\mathcal{H}$. The definition of δ^+ is pictured in Figure 1. It should be remarked that Theorem 3.5 is only true in the C^∞ category; the sets $\mathcal{G}_g$ and $\mathcal{G}_d$ are always invariant under the equivalence relation but the quasi-group of boundary canonical transformations need not act transitively in the analytic category [19].

Theorem 3.5 can be considered as a canonical form theorem, showing that by a boundary-canonical transformation $\Sigma(P)$ can be reduced to the following form near $\sigma \mapsto (0,\tilde{\eta}),\ \tilde{\eta} = (0,\cdots,0,1)$

(3.6) $$\xi^2 + \eta_n^2 \ (\sigma \in \mathcal{E})$$

(3.7) $$\xi^2 - \eta_n^2 \ (\sigma \in \mathcal{H})$$

(3.8) $$\xi^2 - x\eta_n^2 + \eta_1\eta_n \ (\sigma \in \mathcal{G}_d)$$

(3.9) $$\xi^2 + x\eta_n^2 + \eta_1\eta_n \ (\sigma \in \mathcal{G}_g)$$

We remark that in the case $\sigma \in \mathcal{G}_d$, $\partial\chi$ needs only to preserve the 'formal power series' of δ^+ since this is all that is invariantly defined.

Consider the map δ^+ in example (2.3) where it can easily be seen to be just the billiard ball map of Hamiltonian mechanics. Here, $M = \mathbf{R}_t \times X$, and everything is t-translation invariant, so we can ignore the t-component of $(t, \tau) \in T^*\mathbf{R}$ and, to destroy the homogeneity we shall fix $\tau = 1$ as well. Thus, δ^+ will be defined on $T^*\partial X \cong T\partial X$ (∂X being Riemannian) near the image of $\mathcal{G}_g$. $\mathcal{G}$ is simply $S\partial X$, the set of unit tangent vectors to ∂X, $\mathcal{H}$ is its interior and $\mathcal{E}$ its exterior. Then $\delta^+ : \overline{\mathcal{H}}, \mathcal{G}_g \to \overline{\mathcal{H}}, \mathcal{G}_g$ is defined as follows. To $(x, v) \in T\partial X$ with $|v| \leq 1$ associate $w \in T_x X$ with $|w| = 1$, $w = v + \nu$, $\nu \in N_x^+\partial X$. Let $\gamma(t)$ be the geodesic in SX with initial point (x, w) and put $\delta^+(x, v) = (x', v')$ where $\gamma(t') = v' + \nu'$, $\nu' \in N_{x'}\partial X$ and $t' > 0$ is the first point at which $\gamma(t') \in \partial TX$. This construction is pictured in Figure 2, in which one also sees how the formal power series of δ^+ is defined near $\mathcal{G}_d$ (δ^+ itself is defined near $\mathcal{G}_d$ for each given extension of X across ∂X).

Finally in this section let us remark on the complexity which overtakes the equivalence classes in $T^*\partial M\backslash 0$ outside $\mathcal{E} \cup \mathcal{H} \cup \mathcal{G}_g \cup \mathcal{G}_d$. Namely, define

$$\mathcal{G}_i = \{\sigma \in T^*\partial M\backslash 0;\ \text{at}\ \rho \in \iota^{*-1}(\sigma) \cap \Sigma(P),$$
$$\{p, x\} = \{p,\{p, x\}\} = 0,\ \{p,\{p,\{p, x\}\}\}(\rho) \neq 0\}\ .$$

These are points of bicharacteristic inflection. In general, $\mathcal{G}_i$ consists of uncountably many equivalence classes, corresponding to the values of a countable number of (functional) invariants on the surface of maximal degeneracy (see [14]).

4. *Pseudo-differential operators*

Pseudo-differential operators on manifolds with boundary have been considered before — for example the algebra of operators with the transmission property discussed by Boutet de Monvel [2]. The salient property of the class of operators considered here is the way they preserve boundary values.

On $M = \overline{\mathbf{R}_x^+} \times \mathbf{R}_y^n$ by a pseudo-differential operator $A \in L^m(\overline{\mathbf{R}_x^+} \times \mathbf{R}_y^n)$ (always properly supported) we mean an oscillatory integral operator of the form

$$(4.1) \quad Au(x,y) = (2\pi)^{-n-1} \iint_0^\infty e^{i(x-\overline{x})\xi + i(y-\overline{y})\eta}\, a(x,y,\overline{x},\overline{y}, x\xi, \eta)\, u(\overline{x},\overline{y})\, d\overline{x} d\overline{y}\, d\xi\, d\eta \,.$$

Here, $a(x,y,\overline{x},\overline{y},\lambda,\eta)$ is a classical symbol of order m:

$$a \sim \sum_j a_{m-j} \qquad a_{m-j} \text{ homog. degree } m-j \,.$$

The only novelty in (4.1) is the fact that the 'symbol' in the integrand is not a symbol in the usual sense near the boundary because of its special dependence on ξ; this special argument is the key to its behavior. One easily verifies that A defines a continuous linear mapping

$$A : \overset{\circ}{\mathcal{E}}'(\overline{\mathbf{R}^+} \times \mathbf{R}^n) \to \mathcal{E}'(\overline{\mathbf{R}^+} \times \mathbf{R}^n)$$

from the space $\overset{\circ}{\mathcal{E}}' = \{u \in \mathcal{E}'(\mathbf{R} \times \mathbf{R}^n);\ \operatorname{supp} u \subset \overline{\mathbf{R}^+} \times \mathbf{R}^n\}$ supported by $\overline{\mathbf{R}^+} \times \mathbf{R}^n$ to $\mathcal{E}'(\overline{\mathbf{R}^+} \times \mathbf{R}^n) = \mathcal{E}'(\mathbf{R} \times \mathbf{R}^n)|_{\mathbf{R}^+ \times \mathbf{R}^n}$, the space of distributions extendible from $\mathbf{R}^+ \times \mathbf{R}^n$. Indeed, the usual trick of integration by parts allows one to assume that $m << 0$, but this does not immediately give convergence since one only has local estimates

$$|a| \le C(1+|\eta|+|x\xi|)^m$$

so the ξ integral might not converge right up to $x = 0$, it does so after multiplication by x^{-m} (for details of this and subsequent results of the next two sections the reader is referred to the forthcoming paper [15]).

Notice immediately one problem: A does not act on a fixed space of distributions so L cannot form a ring. A related problem is that even when the symbol $a \in S^{-\infty}$ (as a function of λ) it does not necessarily follow that Au is smooth up to $x = 0$ (though certainly $Au \in C^\infty(\mathbf{R}^+ \times \mathbf{R}^n)$). To eliminate both these difficulties we shall restrict the domain of the operators to the space $\mathring{\mathcal{N}}(\overline{\mathbf{R}^+} \times \mathbf{R}^n) \subset \mathring{\mathcal{E}}'(\overline{\mathbf{R}^+} \times \mathbf{R}^n)$ of *normally regular distributions*.

(4.2) DEFINITION. If $u \in \mathring{\mathcal{E}}'(\overline{\mathbf{R}^+} \times \mathbf{R}^n)$ then $u \in \mathring{\mathcal{N}}(\overline{\mathbf{R}^+} \times \mathbf{R}^n)$ if, and only if

1) $Au \in C^\infty(\overline{\mathbf{R}^+} \times \mathbf{R}^n)$ $\forall A \in L^{-\infty}(\overline{\mathbf{R}^+} \times \mathbf{R}^n)$.

2) $Au \in C^\infty([0,\varepsilon) \times \mathbf{R}^n)$, for some $\varepsilon > 0$, if $A \in L^m$ has $a = 0$ in a set $\{(x,y,\bar{x},\bar{y},\lambda,\eta);\ 0 \le x \le \varepsilon',\ 0 \le \bar{x} \le \varepsilon',\ |\lambda| \le \varepsilon'|\eta|\}$ for some $\varepsilon' > 0$.

(4.3) LEMMA. $L^m(\overline{\mathbf{R}^+} \times \mathbf{R}^n)$ *and* $\mathring{\mathcal{N}}(\overline{\mathbf{R}^+} \times \mathbf{R}^n)$, *and* $\mathcal{N}(\overline{\mathbf{R}^+} \times \mathbf{R}^n) = \mathring{\mathcal{N}}(\overline{\mathbf{R}^+} \times \mathbf{R}^n)|_{\mathbf{R}^+ \times \mathbf{R}^n} \subset \mathcal{E}'(\overline{\mathbf{R}^+} \times \mathbf{R}^n)$, *are coordinate free and* $\mathcal{N}(\overline{\mathbf{R}^+} \times \mathbf{R}^n) \subset C^\infty(\overline{\mathbf{R}^+}; \mathcal{E}'(\mathbf{R}^n))$.

As a consequence of this last result, the restriction of elements of $\mathcal{N}(\overline{\mathbf{R}^+} \times \mathbf{R}^n)$ to the boundary, $x = 0$, is well-defined. More generally, we can define $L(M)$, $\mathring{\mathcal{N}}(M)$ and $\mathcal{N}(M)$ for arbitrary manifolds with boundary and then

$$|_{\partial M} : \mathcal{N}(M) \to \mathcal{E}'(\partial M)$$

is surjective. The main reason for considering $\mathcal{N}$ is as a domain for L^m, but it is also directly interesting:

(4.4) LEMMA. *If* $u \in \mathring{\mathcal{E}}'(M)$ *and* $Pu \in \mathring{\mathcal{N}}(M)$ *where* P *is non-characteristic for* ∂M, *then* $u \in \mathring{\mathcal{N}}(M)$.

(4.5) REMARK. $\mathring{\mathcal{N}}(M) = \mathcal{N}(M) \oplus \mathcal{E}'(M, \partial M)$ where $\mathcal{N}(M) \hookrightarrow \mathring{\mathcal{E}}'(M)$ is the inclusion using the fact that $u \in \mathcal{N}(M)$ is smooth (hence locally integrable) in $\mathring{x}$. $\mathcal{E}'(M, \partial M) = \{u \in \mathcal{E}'(M);\ \operatorname{supp} u \subset \partial M\}$.

(4.6) THEOREM. $L(M)$ *acts on* $\mathcal{N}(M)$ *as a filtered ring of operators with the usual symbol properties.*

$$\begin{array}{ccccccc}
0 \hookrightarrow & L^m(M) & \xrightarrow{\sigma_m} & S^m_\partial(S^*M) & \longrightarrow & 0 & \quad \textit{exact} \\
& \uparrow & & & & & \\
0 \hookrightarrow & L^{m-1}(M) & \xrightarrow{\sigma_{m-1}} & S^{m-1}_\partial(S^*M) & \longrightarrow & 0 & \\
& \uparrow & & & & & \\
& \vdots & & & & &
\end{array}$$

$$L^m \circ L^{m'} \subset L^{m+m'}, \quad \sigma_{m+m'}\big|_{L^m \circ L^{m'}} = \sigma_m \cdot \sigma_{m'} .$$

Here, we allow in $L^{-\infty}(M)$ all smoothing operators on $\mathcal{N}(M)$. $S^m_\partial(S^*M)$ is the space of C^∞ functions on $T^*M\backslash 0$ which are homogeneous of degree m and satisfy the trace conditions on ∂T^*M; in (any) local coordinates

$$\partial_x^k b(0, y, \xi, \eta) \text{ is a polynomial of degree } \leq k \text{ in } \xi$$

($\Longleftrightarrow b(x, y, \xi, \eta) = a(x, y, x\xi, \eta)$, indeed for A of the form (4.1) in local coordinates $\sigma_m(A) = a(x, y, x, y, x\xi, \eta)$).

(4.7) THEOREM. *If* $A \in L^m(M)$, $u \in \mathcal{N}(M)$ *then*

$$Au\big|_{\partial M} \equiv (\partial A) u\big|_{\partial M} \pmod{C^\infty}$$

where $\partial A \in L^m(\partial M)$ *is a pseudo-differential operator on* ∂M, $\sigma_m(\partial A) = \sigma_m(A)\big|_{\partial T^*M}$.

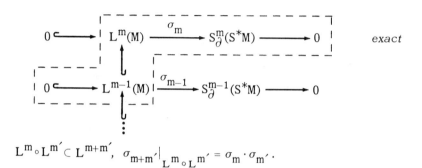

Note that, because of the special form of $\sigma(A)$ its restriction to ∂T^*M $(x=0)$ is constant on the fibers of ι^* (is independent of ξ) so pushes forward to a function on $T^*\partial M\backslash 0$. Indeed, we can consider $\sigma(A)$ as defined on the homogeneous topological space

$$BM = (T^*M\backslash 0)\backslash(N^*\partial M))/\sim \;\cong\; (T^*\mathring{M}\backslash 0) \cup (T^*\partial M\backslash 0)\,,$$

obtained by identifying points in $\partial T^*M\backslash N^*\partial M$ which project onto the same point in $T^*\partial M$.

Now, if $u \in \mathcal{N}(M)$ we can define a notion of singular spectrum (or wave-front set) which has analogues of all the usual properties of SS (see [9]).

(4.8) DEFINITION. If $u \in \mathcal{N}(M)$ then

$$SS_b(u) = BM\backslash\{\rho \in BM;\ \exists A \in L^0(M) \text{ with } \sigma_0(A)(\rho) \neq 0 \text{ and } Au \in C^\infty(M)\}.$$

(4.9) PROPOSITION. $SS_b(u) = SS(u|_{\mathring{M}}) \cup \Delta$, $\Delta = \{\sigma \in T^*\partial M\ 0;$ *in (any) coordinates,* $\sigma = (\bar{y}, \bar{\eta})$, $\exists B \in L^0(\mathbf{R}^n)$ *elliptic at* σ *with* $Bu \in C^\infty([0,\varepsilon)\times \mathbf{R}^n)$ *for some* $\varepsilon > 0\}$.

Thus $SS_b(u)$ corresponds with a natural measure of (failure of) regularity up to the boundary (see [1]).

(4.10 PROPOSITION. $SS(u|_{\partial M}) \subset SS_b(u) \cap (T^*\partial M\backslash 0)$, $SS_b(Au) \subset SS_b(u)$, $\forall A \in L(M)$, $u \in \mathcal{N}(M)$.

It is with respect to this notion of singular spectrum that the singularities of solutions to boundary value problems will be measured.

5. *Fourier integral operators*

The pseudo-differential operators considered in Section 4 are a special case of Fourier integral operators on a manifold with boundary. Each space of F.I.O.'s is associated with a boundary canonical transformation from one manifold to another $\chi : T^*N, \sigma \to T^*M, \sigma'$ where $\sigma \in T^*\partial N\backslash 0$, $\sigma' \in T^*\partial M\backslash 0$ and $\partial\chi(\sigma) = \sigma'$. Thus $L(M)$ is the space of operators associated with the identity transformation on T^*M, (cf. Hörmander [8]; the theory in this section is an extension of Hörmander's, see also [10] for a different class of F.I.O.'s on manifolds with boundary). To define the operators $F : \overset{\circ}{\mathcal{E}}'(N) \to \mathcal{D}'(M)$ associated to χ, $F \in I(N, M; \chi)$, we introduce local coordinates.

Thus, suppose $\chi : T^*(\overline{R^+_x} \times R^n), (0, \eta) \to T^*(\overline{R^+_x} \times R^n), (0, \tilde{\eta})$ where $\tilde{\eta} = (0, \cdots, 0, 1)$. Then, there exists a C^∞ function $\phi(x, y, \overline{y}, \lambda, \theta)$, real-valued and homogeneous of degree 1 in a conic neighborhood Γ of $(0, 0, 0, 0, \tilde{\theta})$, $\tilde{\theta} = (0, \cdots, 0, 1)$ in $\overline{R^+} \times R^n \times R^n \times ((R \times R^N)\backslash 0)$ in which $|\lambda| \le C|\theta|$, such that

(5.1) The $d_{(x,y,y,\theta)}\phi'_{\theta_j}$ are linearly independent in $\phi'_\theta = 0$

(5.2) $$\partial\phi/\partial\lambda \neq 0$$

(5.3) The map $$\mathcal{C}_\phi = \{(x, y, \overline{y}, \xi, \theta);\ \phi'_\theta(x, y, \overline{y}, x\xi, \theta) = 0\} \to (x, y, \phi'_x + \xi\phi'_\lambda, \phi'_y;\ x\phi'_\lambda, \overline{y}, \xi, -\phi'_{\overline{y}})$$

(evaluated at $\lambda = x\xi$) is a local diffeomorphism onto graph χ.

The existence of such a phase function for $x > 0$ follows from the usual parametrization of Lagrangian manifolds (see [8], Guillemin and Sternberg [6]) and the extra conditions on ϕ as $x \downarrow 0$ can be arranged because χ is a boundary canonical transformation. Detailed arguments will appear in [15]. Using ϕ an element $F \in I^m(\overline{R^+_x} \times R^n, \overline{R^+_x} \times R^n; \chi)$ is of the form

(5.4) $$Fu(x, y) = \int \exp(i\,\phi(x,y,\overline{y}, x\xi, \theta) - i\overline{x}\xi)\, a(x,y,\overline{x},\overline{y}, x\xi, \theta)\, u(\overline{x}, \overline{y})\, d\overline{x} d\overline{y}\, d\xi\, d\theta\ ,$$

where, as in Section 4, $a(x, y, \overline{x}, \overline{y}, \lambda, \theta)$ is a classical symbol which has support in a set determined by the domain of ϕ. Following the arguments of [8] one can show that the local definition (5.4) of the space I^m

is coordinate free and independent of the choice of ϕ. Thus I^m can be defined on manifolds with boundary (as before we shall assume all operators to have proper supports).

(5.5) PROPOSITION. $I^m(N, M; \chi)$ *maps* $\mathfrak{N}(N)$ *to* $\mathfrak{N}(M)$, *maps singular spectra*

$$SS_b(Fu) \subset \chi(SS_b(u)) \quad \forall u \in \mathfrak{N}(N)$$

and restricts to the boundary into the space of Fourier integral operators associated to $\partial\chi$

$$Fu|_{\partial M} = \partial F(u|_{\partial N}) \quad (\text{mod. } C^\infty(\partial M)) \tag{5.6}$$

with $\partial F \in I^m(\partial N, \partial M; \partial\chi)$.

The operators in $I^{\cdot}(N,M;_\chi)$ have properties closely analogous to those of Fourier integral operators, including an invariant symbol calculus and composition law leading to Egorov's theorem. This is not surprising since the operators are Fourier integral operators away from the boundary, the proofs of the properties thus reduce to analysis near the boundary itself. The symbol map σ_m gives a diagram as in Theorem 4.6 except that the appropriate symbol space, in the case discussed above with χ a boundary canonical transformation, will be the space $S^m_\partial(S^*M; L)$ of homogeneous sections, satisfying the trace conditions, of the Maslov bundle defined by χ.

If $F \in I^m(N,M;_\chi)$, $F' \in I^{m'}(M,M';_{\chi'})$. $F' \circ F \in I^{m+m'}(N,M';_{\chi' \circ \chi})$ and

$$\sigma_{m+m'}(F' \circ F) = \chi^* \sigma_{m'}(F') \cdot \sigma(F) .$$

The symbols are well-behaved under the restriction map $F \mapsto \partial F$ in (5.6) with

$$\sigma_m(\partial F) = \sigma_m(F)|_{S^*(\partial M)} ,$$

this restriction being meaningful since the sections in $S^m_\partial(S^*M; L)$ can actually be considered as sections of a bundle over BM.

6. *Normal forms*

Using the calculus of operators discussed briefly above and the geometrical result Theorem 3.5 one can, microlocally, reduce a given operator P to normal form near $\sigma_0 \in T^*\partial M\backslash 0$, if σ_0 is non-degenerate.

(6.1) THEOREM. *If* $\sigma_0 \in \mathcal{E}$ *(resp.* $\mathcal{H}, \mathcal{G}_d, \mathcal{G}_g$*) for* P *on* M $\exists F, G \in \overset{\circ}{I}(M, \mathbf{R}^+_x \times \mathbf{R}^n_y; \chi)$ *such that*

(6.2) $$SS_b(P'Fu - GP_u) \not\ni (0, \tilde{\eta}) \quad \forall u \in \mathcal{N}(M)$$

(6.3) $$F \text{ is elliptic at } (\sigma_0, (0, \tilde{\eta}))\,.$$

Here, P′ *is the appropriate normal form at* $(0, \tilde{\eta}), \tilde{\eta} = (0, \cdots, 0, 1)$.

(6.4) $$P' = D_x^2 + D_{y_n}^2 \qquad (0, \tilde{\eta}) \in \mathcal{E}$$

(6.5) $$P' = D_x^2 - D_{y_n}^2 \qquad (0, \tilde{\eta}) \in \mathcal{H}$$

(6.6) $$P' = D_x^2 - xD_{y_n}^2 + D_{y_1} D_{y_n} \qquad (0, \tilde{\eta}) \in \mathcal{G}_d$$

(6.7) $$P' = D_x^2 + xD_{y_n}^2 + D_{y_1} D_{y_n} \qquad (0, \tilde{\eta}) \in \mathcal{G}_g\,.$$

Of course, if $Pu \in C^\infty(M)$ and $\sigma_0 \notin SS(u|_{\partial M})$ where σ_0 lies in $\mathcal{E}, \mathcal{H}$, $\mathcal{G}_d$ or $\mathcal{G}_g$ then $\sigma_0 \notin SS_b(P'Fu)$ and $Fu|_{x=0} \equiv \partial F(u|_{\partial M})$ with $\partial F \in \overset{\circ}{I}(\partial M, \mathbf{R}^n; \partial\chi)$ so $\partial\chi(\sigma_0) = (0, \tilde{\eta}) \notin SS(Fu|_{x=0})$. Thus, to analyze the singularities of u it suffices to discuss the operators P′ microlocally. This is relatively straightforward since all the P′ have coefficients independent of the y-variables.

(6.8) PROPOSITION ($\mathcal{E}$). *If* $P' = D_x^2 + D_{y_n}^2$, $u \in \mathcal{N}(\overline{\mathbf{R}^+} \times \mathbf{R}^n)$, $(0,\tilde{\eta}) \notin SS(u|_{x=0})$ *and* $(0,\tilde{\eta}) \notin SS_b(P'u)$ *then* $(0,\tilde{\eta}) \notin SS_b(u)$.

Thus, if $Pu \in C^\infty(M)$ and $u|_{\partial M} \in C^\infty(\partial M)$ then $SS_b(u) \cap \mathcal{E} = \emptyset$. We can easily reinterpret this in a useful way. First, let $\Sigma \subset T^*M\backslash 0$ be the characteristic surface of P, defined by $p = 0$, Σ projects into BM to the closed conic set which we shall denote Σ_b. Thus, $\Sigma_b \cap T^*\mathring{M}\backslash 0 = \Sigma \cap T^*\mathring{M}\backslash 0$ and by considering the coordinates of Lemma 1.6 one easily sees that $\Sigma_b \cap (T^*\partial M\backslash 0) = \mathcal{H} \cup \mathcal{G}$. Thus, Proposition 6.8 and the discussion above give

(6.9) THEOREM. *If* P *is of real principal type with* ∂M *non-characteristic,* $Pu \in C^\infty(M)$ *and* $u|_{\partial M} \in C^\infty(\partial M)$ *then* $SS_b(u) \subset \Sigma_b$.

Next consider the hyperbolic case: $P' = D_x^2 - D_{y_n}^2$, now there are two points $(0, 0, \pm 1, \tilde{\eta})$ in Σ above $(0, \tilde{\eta})$ and through each of these points there is a bicharacteristic, half of which lies over $x \geq 0$ (see Figure 3). For $p' = \xi^2 - \eta_n^2$ the bicharacteristic equations are

$$\frac{dx}{dt} = 2\xi, \quad \frac{dy_n}{dt} = -2\eta_n$$

with everything else constant. Thus the two bicharacteristics have $x = \pm y_n, y_1 = \cdots = y_{n-1} = 0$, $\eta_1 = \cdots = \eta_{n-1} = 0$, $\xi = \mp \eta_n$.

(6.10) PROPOSITION. *If* $P' = D_x^2 - D_{y_n}^2$, $u \in \mathcal{N}(\overline{\mathbf{R}^+} \times \mathbf{R}^n)$, $(0,\tilde{\eta}) \notin SS_b(P'u)$ *and* $(0,\tilde{\eta}) \notin SS(u|_{x=0})$ *then either both or neither of the two half bicharacteristics above* $(0,\tilde{\eta})$ *lie (locally) in* $SS_b(u)$.

In general then, in $\Sigma_b \cap (T^*\mathring{M}\backslash 0 \cup \mathcal{H})$ there is a nice *broken bicharacteristic flow* $F : D \to (\Sigma_b\backslash \mathcal{G})$ where D is open, $D \supset \{0\} \times (\Sigma_b\backslash\mathcal{G})$ and is maximal with

$$F(x, F(t,\rho)) = F(s+t,\rho) \text{ if } (t,\rho) \in D, (s, F(t,\rho)) \in D .$$

Then Proposition (6.10) and the transformation theory above give:

(6.11) THEOREM. *If* P *is of real-principal type, with* ∂M *non-characteristic,* $Pu \in C^\infty(M)$ *and* $u|_{\partial M} \in C^\infty(\partial M)$ *then* $SS_b(u) \subset \Sigma_b$ *is invariant under the broken bicharacteristic flow in* $\Sigma_b \setminus \mathcal{G}$ *(i.e.,* $SS_b \cap (\Sigma_b \setminus \mathcal{G})$ *is the union of broken bicharacteristics).*

The results near $\mathcal{G}_d$ and $\mathcal{G}_g$ can be expressed in the same form, as can the more general results discussed in §8. More generally still, if one just assumes that $Pu \in C^\infty(M)$ then $SS_b(u) \setminus SS(u|_{\partial M}) \subset \Sigma_b$ and is invariant under the broken Hamilton flow in $\Sigma_b \setminus (\mathcal{G} \cup SS(u|_{\partial M}))$, in the usual sense of restriction of flows.

Define $\Sigma_b^{(3)} = \mathcal{G} \setminus (\mathcal{G}_g \cup \mathcal{G}_d)$, (part of the more general notation introduced in [16], and §8).

(6.12) THEOREM. *The broken Hamilton flow extends by continuity to* $\Sigma_b \setminus \Sigma_b^{(3)}$ *and if* $Pu \in C^\infty(M)$ *then* $SS_b(u) \setminus SS(u|_{\partial M}) \subset \Sigma_b$ *is invariant under the flow in* $SS_b(u) \setminus (SS(u|_{\partial M}) \cup \Sigma_b^{(3)})$.

This theorem summarizes results obtained by Chazarain ([3], for $\mathcal{E}$ and $\mathcal{H}$), Lax-Nirenberg ([18], for $\mathcal{E}$ and $\mathcal{H}$), Melrose ([11] for $\mathcal{G}_d$), Taylor ([20], for $\mathcal{G}_d$ when P is hyperbolic), Andersson-Melrose ([1], for $\mathcal{G}_g$) and Eskin ([4], for $\mathcal{G}_g$ when P is hyperbolic), using a variety of methods, all different from the transformation approach outlined above.

Returning to the proof of Theorem 6.12, we note again that it suffices to prove it, microlocally for the two simple operators P' in (6.6), (6.7). This is not completely trivial; let us consider $\mathcal{G}_d$. The result for a simple operator, essentially P' in (6.6), was first proved by Friedlander [5]. The method is to construct, directly, solutions to

$$(6.13)\qquad \begin{aligned} D_x^2 u - xD_{y_n}^2 u + D_{y_1} D_{y_n} u &= 0 \\ u|_{x=0} &\equiv u_0 \pmod{C^\infty} \end{aligned}$$

where $u_0 \in \mathcal{E}'(\mathbf{R}^n)$ has its singular spectrum near $(0, \tilde{\eta})$. By partial Fourier transformation (formally)

$$\hat{u}(x, \eta) = (2\pi)^{-n} \int e^{-iy\cdot\eta} u(x, y) \, dy$$

(6.13) becomes

$$\frac{d^2 \hat{u}}{dx^2} = (x\eta_n^2 - \eta_1 \eta_n) \hat{u} .$$

Changing to the variable $\zeta = (x\eta_n^2 - \eta_1\eta_n)\eta_n^{-4/3}$ this is Airy's equation for $v(\zeta) = \hat{u}(x, \eta)$

$$\frac{d^2}{d\zeta^2} v(\zeta) = \zeta \, v(\zeta) .$$

All Airy functions are entire and can be expressed in terms of the one basic function

$$\text{Ai}(z) = \int_\gamma \exp(zs - s^3/3) \, ds \tag{6.14}$$

where γ is a contour as in Figure 5. The bicharacteristics of P' near $(0, \eta)$ are either glancing rays as in Figure 4 or reflected rays as in Figure 3. By choosing the appropriate Airy functions $A_\pm(z) = \text{Ai}(e^{\pm 2\pi i/3} z)$ (corresponding to (6.14) with γ replaced by $\gamma_\pm$ in Figure 5), one finds that

$$\hat{u}_\pm(x, \eta) = \hat{u}_0(\eta) \rho(\eta) \frac{A_\pm(\zeta(x, \eta))}{A_\pm(\zeta(0, \eta))} ,$$

with ρ a homogeneous cut-off factor localizing near $\tilde{\eta}$, provides solutions of (6.13) by inverse Fourier transformation. One solution has its singular spectrum 'forwards' along the bicharacteristics and the other 'backwards.' By duality arguments these solutions can be shown to be microlocally unique and using them one easily proves the part of Theorem 6.12 concerned with $\mathcal{G}_d$.

The analysis of points in $\mathcal{G}_g$ is similar, since (6.7) again reduces to the Airy operator upon Fourier transformation. The bicharacteristics of P' in (6.7), near $(0, \tilde{\eta})$, look like Figure 6 (just Figure 4 reversed about the boundary, with reflections). The limiting ray along which singularities from $(0, \tilde{\eta})$ can propagate is the curve $x = 0$, $\xi = 0$, $y_1, \cdots, y_{n-1} = 0$, $\eta_1 = \cdots = \eta_{n-1} = 0$, $\eta_n = 1$, $y_n = t$, a 'boundary bicharacteristic,' projecting to a *gliding ray* in $T^*\mathbf{R}^n$ (see [1]).

7. *Other boundary conditions*

To analyze the singularities when u is subject to more general boundary conditions (1.9), one can use the results of §6 above to analyze the *Neumann map*. The reduction to normal form shows that, if $\sigma_0 \in \Sigma_b \setminus \Sigma_b^{(3)}$ then the conditions:

(7.1) $$\sigma_0 \notin SS_b(Pu), \ \sigma_0 \notin SS(u|_{\partial M} - u_0)$$

(7.2) $SS_b(u)$ does not contain the backward half bicharacteristics from points near σ_0

(where (7.2) is void if $\sigma_0 \in \mathcal{E}$) determine $u \in \mathcal{N}$ microlocally from u_0, i.e. any other solution $u' \in \mathcal{N}$ satisfies $\sigma_0 \notin SS_b(u' - u)$. In particular, if ∂_ν is the Neumann vector field (defined by the condition that it be $\frac{\partial}{\partial x}$ in the coordinates of Lemma 1.6) then $N_+ u_0 = \partial_\nu u|_{\partial M}$ is a well-defined microlocal operator on ∂M, the forward Neumann map (similarly one defines the 'backwards' Neumann map by replacing backward by forward in (7.2); for $\sigma_0 \in \mathcal{E}$ $N_+ \equiv N_-$).

One easily sees that u_0 and $N_+ u_0$ together determine u satisfying (7.1), (7.2) microlocally near σ_0, and the singularities of u can be deduced from those of u_0, $N_+ u_0$. Now, suppose that B is a first-order operator, without loss of generality in (1.9) we can assume

$$B = \partial_\nu + T$$

where T is a first-order (pseudo-) differential operator on ∂M. To analyze the singularities of u, given that $Bu \in C^\infty$, one examines the solutions $u_0, u_1 \in \mathcal{E}'(\partial M)$ to

$$
(7.3) \qquad \begin{aligned} u_1 + T u_0 &= 0 \\ u_1 - N_\pm u_0 &= 0 \end{aligned}
$$

as microlocal equations at σ_0. If, for example, these have no solutions with $\sigma_0 \in SS(u_0)$, for both signs, one easily deduces the appropriate analogue of Theorem 6.17, microlocally at σ_0. That is, $SS_b(u)\backslash(SS(Bu|_{\partial M}))$ is invariant under the broken Hamilton flow near σ_0.

Now, near $\sigma_0 \in \mathcal{E} \cup \mathcal{H}$, $N_\pm$ are (classical) pseudo-differential operators and a sufficient condition to guarantee the result above is that the system (7.3), or just the operators $N_\pm + T$, be elliptic at σ_0. This is the *Lopatinskii-Schapiro* condition. Of course, the behavior of solutions to (7.3), and hence the behavior of u, can be discussed more generally than this (see for example [21]). Near $\sigma_0 \in \mathcal{G}_d \cup \mathcal{G}_g = \Sigma_b^2$, $N_\pm$ are not so simple. Near $\sigma_0 \in \mathcal{G}_d$ they are 'Airy operators of the first kind' which are pseudo-differential operators, but not in the standard coordinate free classes, (see [13]). Nevertheless, one can analyze (7.3) and one finds hypoellipticity at $\sigma_0, \sigma_0 \notin SS(u_0)$, not only under the Lopatinskii-Schapiro condition, which is just $\sigma_1(T) \neq 0$ at σ_0, but also under the generalized Neumann condition $\sigma_1(T) \equiv 0$ on $\mathcal{G}_d$. This is due, in essence, to the fact that $N_\pm$ have symbols which are the 'square roots' of the function $r_0(y,\eta)$, which vanishes simply on $\mathcal{G}$.

Similar results can be expected to hold near points of $\mathcal{G}_g$ (and propagation of singularities results are known by a different method: [1]) but the operators $N_\pm$ are more complicated, not being microlocal.

8. *Higher-order tangency*

The results of this section were obtained jointly with J. Sjöstrand, details can be found in [16].

(8.1) DEFINITION. For $k \geq 3$ let $\Sigma_b^k \subset \mathcal{G}$ be the set of points at which bicharacteristics have exactly k^{th} order contact with the boundary.

Recall that if $\sigma \in \mathcal{G}$ $\exists!$ $\rho \in \Sigma$ with $\iota^*(\rho) = \sigma$. The definition of $\sigma \in \Sigma_b^k$ says simply

$$H_p^j x = 0 \text{ at } \rho, \ j = 0, \cdots, k-1, \ H_p^k x \neq 0 \text{ at } \rho .$$

Of course, since $\sigma \in \mathcal{G}$, $x = 0$ and $H_p x = 2\xi = 0$ at ρ. Working in the coordinates of Lemma 1.6 we find

(8.2) LEMMA. $\Sigma_b^k = \{\sigma = (y, \eta) \in T^*\partial M \backslash 0;\ r_0(y, \eta) = r(0, y, \eta) = 0$ *and* $H_{r_0}^j r_1(y, \eta) = 0,\ j = 0, \cdots, k-3,\ H_{r_0}^{k-2} r_1(y, \eta) \neq 0$ *where* $r_1(y, \eta) = r'_x(0, y, \eta)\}$, *for* $k \geq 2$.

Put $\Sigma_b^\infty = \{(y, \eta);\ r_0 = H_{r_0}^j r_1 = 0\ \forall j\}$. $\Sigma_b^{(k)} = \bigcup_{\infty \geq j \geq k} \Sigma_b^j$. Then $\Sigma_b^{(k)}$ is closed for each k, $\Sigma_b^{(2)} = \mathcal{G}$, $\Sigma_b^{(1)} = \mathcal{G} \cup \mathcal{H}$ and Σ_b^∞ is the set of points at which the bicharacteristics are tangent to the boundary to infinite order.

(8.3) THEOREM. *If* P *is of real-principal type and* ∂M *is non-characteristic, the broken Hamilton flow extends by continuity to a flow on* $\Sigma_b \backslash \Sigma_b^\infty$ *and, if* $Pu \in C^\infty(M)$, $SS_b(u) \backslash SS(u|_{\partial M})$ *is invariant under it.*

The proof of this theorem, which is microlocal in nature, is carried out by induction over k, the order of tangency, starting with the results of §6 for $\Sigma^0 = \Sigma_b \cap T^*M$, $\Sigma_b^1 = \mathcal{H}$ and $\Sigma_b^2 = \mathcal{G}_d \cup \mathcal{G}_g$. Consider first a point $\sigma_0 \in \Sigma_b^k \subset \mathcal{G}$ $(k \geq 3)$. (See Figure 7.) Near σ_0, $\Sigma_b^{(k)}$ is defined by $H_{r_0}^j r_1 = 0,\ j = 0, \cdots, k-3$. In $\Sigma_b^{(k)}$, Σ_b^k is defined by $H_{r_0}^{k-2} r_1 \neq 0$. So $H_{r_0}(f_k) \neq 0$ where $f_k = H_{r_0}^{k-3} r_1$ and, near σ_0, $\Sigma_b^{(k)} = \Sigma_b^k \subset V_k$ where $V_k \subset \mathcal{G}$ is the non-singular hypersurface transversal to $H_{r_0}: V_k = \{f_k = 0\}$.

In general the set, $\Sigma_b^{(3)}$, of points we have not yet analyzed can be a complicated C^∞-variety in $\mathcal{G}$. But this observation allows us to use induction over k. Thus, suppose that Theorem (8.3) has been established with Σ^∞ replaced by $\Sigma^{(k)}$ for some k. To estend to the next value, $k+1$, we must show that if $\sigma \in \Sigma_b^k$, $\sigma \notin SS(u|_{\partial M})$ then either the broken bicharacteristic through σ is (locally) contained in $SS_b(u)$ or it is disjoint from it.

Now, let the flow be denoted by $F(\sigma, t)$, so the ray through σ_0 is $t \mapsto F(\sigma_0, t)$. We shall divide $\Sigma^{(2)} \backslash \Sigma^\infty$ into two regions:

$$\Sigma^{(2),+} = \{\sigma \in \Sigma^{(2)} \backslash \Sigma^\infty;\ \text{there is a free ray (i.e. bicharacteristic) through } \sigma \text{ in } t \geq 0\}$$

and $\Sigma_b^{(2),-} = \Sigma_b^{(2)} \backslash (\Sigma_b^{(2),+} \cup \Sigma_b^\infty)$. The ray through $\sigma \in \Sigma_b^{(2),+}$ is the free ray in $t \geq 0$, for t small. Through a point in $\Sigma^{(2),-}$ the forward ray is the gliding ray–the integral curve of H_{r_0} in $\mathcal{G}$ through σ. Points in Σ^∞ may have both a gliding ray and an infinitely reflected ray (see the example in [20]) in the forward direction. For this reason F does not extend as a flow to Σ_b, in general. Nevertheless, results quite similar to Theorem 8.3 hold even near Σ_b^∞ (see [17]).

From direct estimation of the bicharacteristic equations it follows that, not only does the ray through $\sigma \in \Sigma_b^k$ immediately enter $\Sigma_b \backslash \Sigma_b^{(3)}$ (from the transversality of H_{r_0} to V_k it follows that the ray enters $\Sigma_b \backslash \Sigma_b^{(k)}$) but that the ray is not too close to $\Sigma_b^{(k)}$ for $t \neq 0$, $k \geq 3$. Thus, to carry through the inductive step some method for deducing that $\sigma \notin SS_b(u)$ when we know $\sigma \notin SS(u|_{\partial M})$ and that $\sigma' \notin SS_b(u)$ for σ' 'near' $F(\sigma, t)$, $t < 0$, is needed. To do this we use a commutation argument which will only be briefly outlined here.

For $u, v \in \mathcal{D}'(\overline{\mathbf{R}^+} \times \mathbf{R}^n)$ put

$$< u, v > = \iint_0^\infty u(x,y)\, v(x,y)\, dx\, dy$$

(when this makes any reasonable sense). Then, if $u|_{x=0} = 0$, $P = D_x^2 + R(x, y, D_y)$ and $Q = Q(x, y, D_y)$ is a compactly supported pseudo-differential operator

$$<([P, Q] + (R^* - R)Q)u, u> = <Qu, Pu> - <QPu, u>, \tag{8.4}$$

by integration by parts, the boundary terms all being zero. Now, $Pu \in C^\infty$ so both terms on the right in (8.4) make sense. Suppose we could choose Q so that the operator

$$\frac{1}{i} N = [P, Q] + (R^* - R)Q = A^*A \tag{8.5}$$

where A is a pseudo-differential operator in the y-variables elliptic at σ, then we would deduce that $\|Au\|^2_{L_2} < \infty$, from which one could conclude that $\sigma \notin SS_b(u)$. Now, in [16], (8.5) is not quite achieved, this failure being seen in the appearance of extra terms on the right. These terms are of four types. First there is an extra positive term $A_1^*A_1$, which is actually a pseudo-differential operator in all variables, but has its support away from $x = 0$. Secondly there are non-positive terms which are bounded on u in view of the assumptions on $SS_b(u)$ near the backward ray through σ. Thirdly there is a term which is bounded because u is smooth away from the characteristic set of P and finally there are lower order terms, which necessitate an inductive argument over orders of Sobolev regularity. Moreover the situation is complicated by the need to approximate Q by smoothing operators so that (8.4) is meaningful. Nevertheless the argument goes through much as indicated. Ignoring all the details let us just consider a suitable choice for the symbol of Q (we shall further simplify the discussion by ignoring the requirement of homogeneity).

If $Q(x, y, D_y)$ is a pseudo-differential operator with symbol $q(x, y, \eta)$ then $[P, Q]$ is an x-differential, y-pseudo-differential operator with symbol

$$\sigma([P, Q]) = i H_p q .$$

We shall assume that q is independent of x for small x and then $[P, Q]$

is a pseudo-differential operator in y, for small x, with symbol $i\,H_r q$. To see a good choice for q, introduce coordinates $(x, t, s_1, \cdots, s_{2n-1})$ in place of $x, y_1, \cdots, y_n, \eta_1, \cdots, \eta_n$ such that $H_{r_0} = \partial/\partial t$. Then,

$$H_r = \frac{\partial}{\partial t} + O(x)\, V\left(\frac{\partial}{\partial t}, \frac{\partial}{\partial s}\right),$$

V a smooth vector field. We choose

$$q(x, t, s) = \beta\left(\frac{t}{\varepsilon^2}\right) \chi\left(\frac{t}{\delta\varepsilon} + \frac{s^2}{\varepsilon^4} + f\left(\frac{x}{\varepsilon^2}\right)\right)$$

where χ, f, β are three C^∞ functions, all real-valued, with graphs as pictured in Figure 8, and $\delta, \varepsilon > 0$ are parameters, δ will be chosen small, depending on the coefficients of p, etc., whereas $\varepsilon < \varepsilon_0$ will be a small parameter needed to guarantee boundedness of the terms supported near the ray backwards through σ. The support of q is a truncated parabolic neighborhood of the gliding ray through 0, see Figure 9. In supp q $x \le \varepsilon^2$, $|s| \le \varepsilon^2$, $-\varepsilon^2 \le t \le \delta\varepsilon$, and q is independent of x in $0 \le x \le \varepsilon^2/2$. We can take σ_0 to be the point $t = \frac{\delta\varepsilon}{2}$, $s = 0$, $x = 0$.

Now,

$$\begin{aligned} -H_r q &= -\left(\frac{\partial q}{\partial t} + O(x)\, V_q\right) \\ &= -\left(\frac{1}{\delta\varepsilon} + O(\varepsilon^2)\right) \cdot \left(O\left(\frac{1}{\delta\varepsilon}\right) + O\left(\frac{|s|}{\varepsilon^4}\right)\right) \chi' \qquad (8.6) \\ &= -\left(\frac{1}{\delta\varepsilon} + O(1)\right) \chi'. \end{aligned}$$

This gives the desired positivity, since we assume χ to be decreasing. Moreover the second term in (8.4) contributes a term $O(1) \cdot \chi$ and we can assume $|\chi| \le |\chi'|$ so formally q has the right behavior in $x \le \frac{\varepsilon^2}{2}$. In $\frac{\varepsilon^2}{2} \le x \le \varepsilon^2$ the extra term $-2\xi \frac{\partial}{\partial x} q$ appears in the symbol of N, thus is bounded by

$$O\left(\frac{1}{\varepsilon^2}\right) \cdot |\xi| \cdot \chi'$$

which is only negligible compared to (8.6) when $|\xi| \leq \varepsilon$. Fortunately this holds in a neighborhood of the characteristic set, so choosing $\delta > 0$ small one gets the desired positivity of the symbol of N, modulo the error terms mentioned above.

9. *Example (2.3) again*

The results discussed in §§6 and 8 can be applied to find the wave fronts of the 'fundamental solution' of the boundary problem of Example 2.3. In Figure 10, with X the complement of the region pictured in Figure 2, the position of the wave fronts at various times $t_1, t_2, \cdots$ is indicated, for the solution with initial, Dirac, singularity at x_0:

$$(\partial_t^2 - \Delta)u = 0, \qquad u|_{R_t \times \partial X} = 0$$

$$u|_{t=0} = \delta_{x_0}(x), \qquad \partial_t u|_{t=0} = 0 .$$

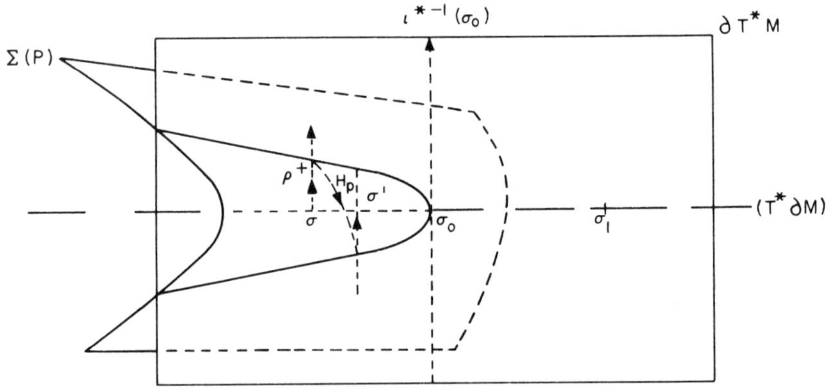

FIGURE 1

$\sigma_0 \in \mathcal{G}_g$ if T^*M is in front of ∂T^*M ,

$\sigma_0 \in \mathcal{G}_d$ if T^*M is behind ∂T^*M ,

$\sigma_1 \in \mathcal{E}$, $\sigma \in \mathcal{H}$ with $\sigma' = \delta^+(\sigma)$.

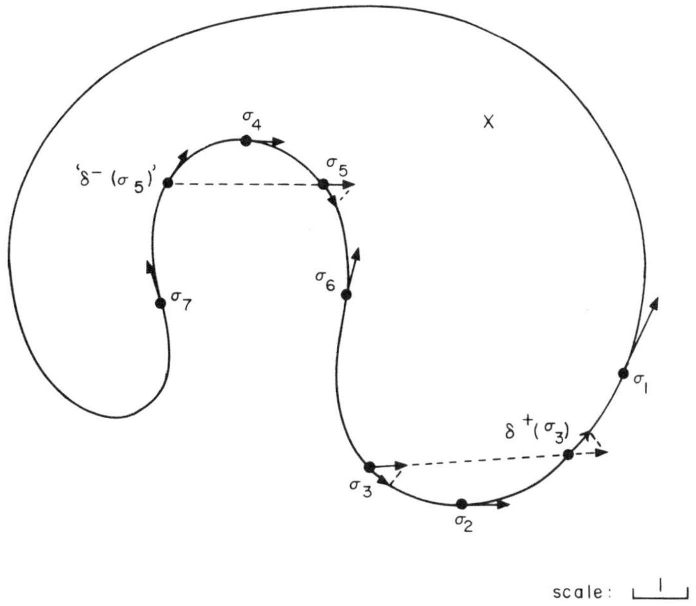

FIGURE 2

$\sigma_1 \in \mathcal{E}$

$\sigma_2 \in \mathcal{G}_g$

$\sigma_3,\ \delta^+(\sigma_3) \in \mathcal{H}$

$\sigma_4 \in \mathcal{G}_d$

$\sigma_5,\ '\delta^-(\sigma_5)' \in \mathcal{H}$

$\sigma_6,\ \sigma_7 \in \mathcal{G}_i \subset \mathcal{G} \backslash (\mathcal{G}_d \cup \mathcal{G}_g)$

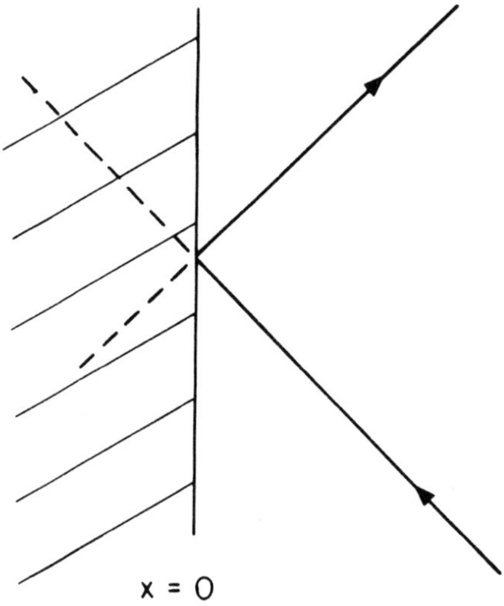

FIGURE 3

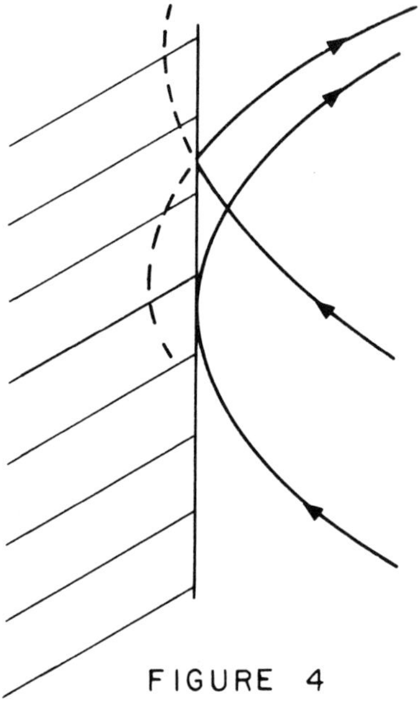

FIGURE 4

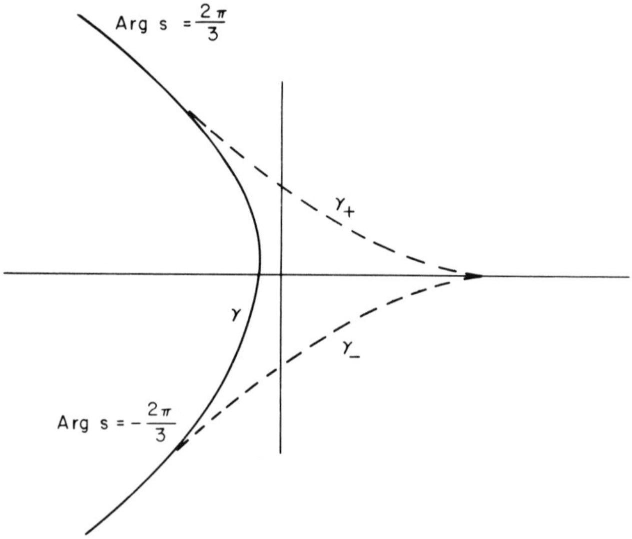

FIGURE 5

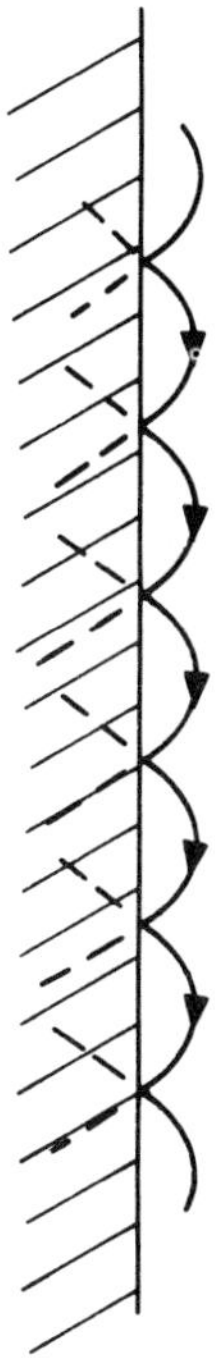

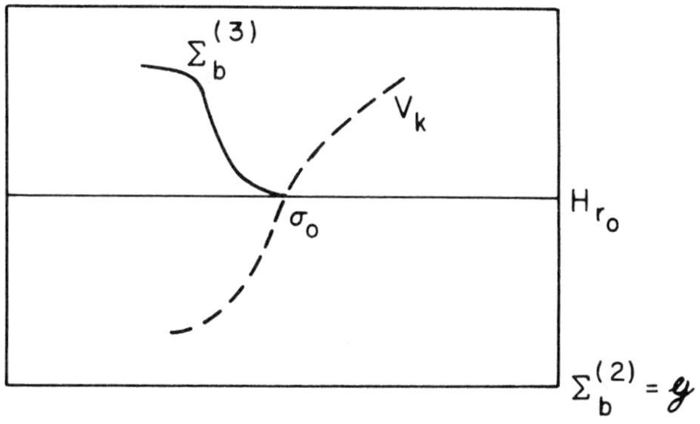

FIGURE 7

FIGURE 6

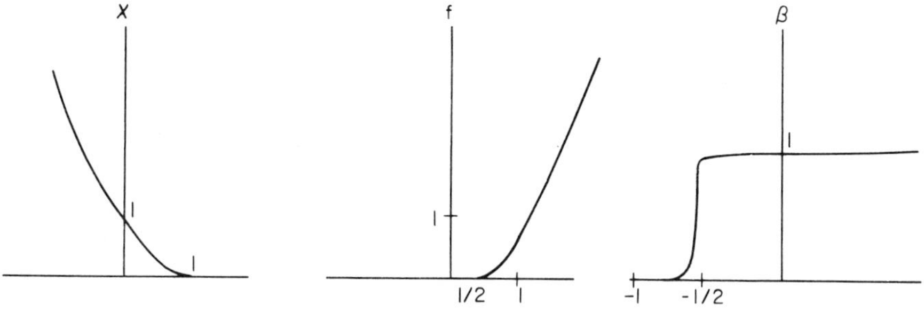

FIGURE 8

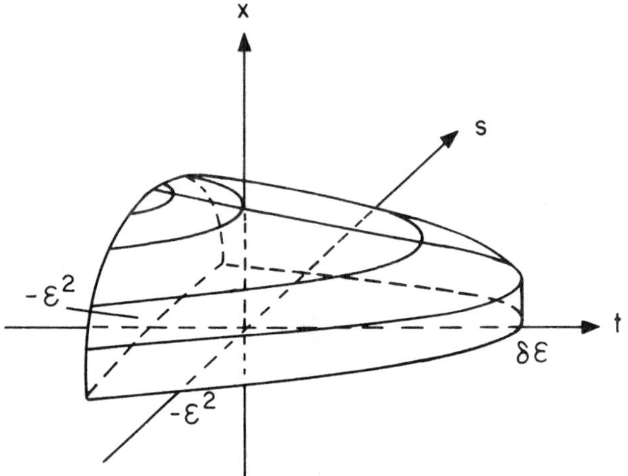

FIGURE 9

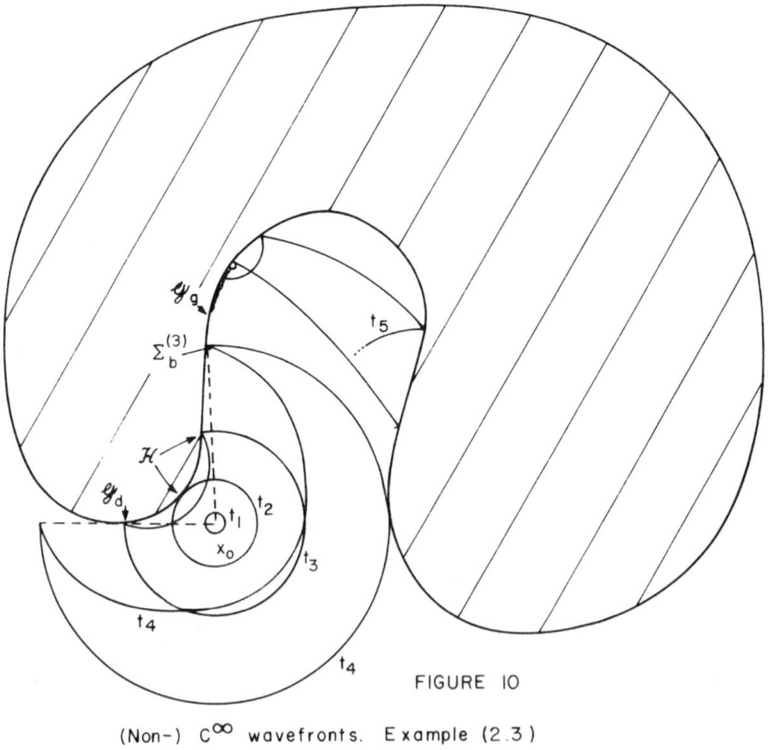

FIGURE 10

(Non-) C^∞ wavefronts. Example (2.3)

Only part of the wavefront is shown for t_5.

REFERENCES

[1] Andersson, K. G., and R. B. Melrose, The Propagation of singularities along gliding rays, Invent. Math., 41, 197-232 (1977).

[2] Boutet de Monvel, L., Boundary problems for pseudodifferential operators, Acta Math., 126, 11-51 (1971).

[3] Chazarain, J., Construction de la parametrix du problème mixte hyperbolique pour l'équation des ondes, C. R. Acad. Sci., Paris 276, 1213-1215 (1973).

[4] Eskin, G., A parametrix for interior mixed problems for hyperbolic equations, J. Anal. Math., 32, 17-62 (1977).

[5] Friedlander, F. G., The wavefront set of a simple initial-boundary value problem with glancing rays, Math. Proc. Camb. Phil. Soc. 79, 145-149 (1976).

[6] Guillemin, V., and S. Sternberg, Geometric Asymptotics, Mathematical Surveys no. 14, A.M.S., Providence, 1977.

[7] Hörmander, L., Linear partial differential operators, Berlin-Heidelberg-New York, Springer, 1963.

[8] ————, Fourier integral operators, I, Acta Math., 127, 179-183 (1971).

[9] ————, Spectral analysis of singularities, Seminar on Singularities of Solutions of Linear Partial Differential Equations, Princeton Univ. Press, Princeton, 1978.

[10] Melin, A., and J. Sjöstrand, A calculus for Fourier integral operators in domains with boundary and applications to the oblique derivative problem, Comm. in P.D.E., 2, 857-935 (1977).

[11] Melrose, R. B., Microlocal parametrices for diffractive boundary value problems, Duke Math. J. 42, 605-635 (1975).

[12] ————, Equivalence of glancing hypersurfaces, Invent. Math. 37, 165-191 (1976).

[13] ————, Airy operators, Comm. in P.D.E., 3, 1-76 (1978).

[14] ————, Equivalence of glancing hypersurfaces, II, to appear.

[15] ————, Transformation of boundary value problems, in preparation.

[16] Melrose, R. B., and J. Sjöstrand, Singularities of boundary value problems, I, Comm. Pure Appl. Math., 31, 593-617 (1978).

[17] ————, II, in preparation.

[18] Nirenberg, L., Lectures on linear partial differential equations, A.M.S. Regional Conference Series, No. 17 (1973).

[19] Oshima, T., Sato's problem in contact geometry, (1976).

[20] Taylor, M. E., Grazing rays and reflection of singularities to wave equations, Comm. Pure Appl. Math. 29, 1-38 (1976).

[21] ———, Rayleigh waves for the equations of linear elasticity, to appear.

PROPAGATION OF SINGULARITIES FOR A CLASS OF OPERATORS WITH DOUBLE CHARACTERISTICS

Nicholas Hanges*

0. *Introduction*

In this lecture we will study pseudo-differential operators P with real principal symbol. It is assumed that the characteristic variety of P is the union of two smooth hypersurfaces with noninvolutive intersection. The construction of microlocal parametrices enables us to study the propagation of singularities.

The approach outlined here is that of Hanges [8], and we refer to that paper for complete details. Working independently, Ivrii [11], [12] and Melrose [14], [15] have obtained essentially the same results as us. Also see Alinhac [0], for the study of a closely related system, Kashiwara, Kawai and Oshima [19], [20] (or Miwa [17]) for a more general discussion in the real analytic case, and Ivrii [13] for some more recent results.

We believe that the explicit formulae of sections 4 and 5 provide a short, direct approach that will be useful in other multiple characteristic problems.

1. *Statements of results*

Notation used and not defined here is standard. See Hörmander [9] and Duistermaat-Hörmander [4].

---

*Supported in part by NSF grant MCS77-18723.

Seminar on Singularities of Solutions
0-691-08221-9/78/0113 $00.70/1 (cloth)
0-691-08213-8/78/0113 $00.70/1 (paperback)

Let $X \subset \mathbf{R}^{n+1}$ be an open set. By $L_c^m(X)$ we mean the collection of all $A \in L_{1,0}^m(X)$ whose full symbol has an asymptotic expansion in positively homogeneous terms of orders $m, m-1, m-2, \cdots$. Let $P_j \in L_c^{m_j}(X)$ for $j = 1, 2$ and $Q \in L_c^{m_1+m_2-1}(X)$, with principal p_j and q. Assume the p_j are real valued.

We will be studying pseudo-differential operators

$$P = P_1 P_2 + Q \tag{1.1}$$

near noninvolutive double characteristics. Precisely speaking, we will study P near $z_0 \in T^*(X)\backslash 0$ where

$$p_1(z_0) = p_2(z_0) = 0 \tag{1.2}$$

$$\{p_1, p_2\}(z_0) \neq 0 . \tag{1.3}$$

(Recall that $\{p_1, p_2\}$ is the Poisson bracket of p_1 and p_2.)

It follows from (1.3) that the Hamilton fields H_{p_1}, H_{p_2} and the cone axis are linearly independent at z_0. In particular, there exists an open interval $O \subset \mathbf{R}$, with $0 \in O$ such that γ_j (the bicharacteristic strip of p_j through z_0) restricted to O is an injective map:

$$\gamma_j : O \to \Sigma_j = \{z \in T^*(X)\backslash 0 : p_j(z) = 0\}$$

$$\gamma_j(0) = z_0$$

for $j = 1, 2$.

Note that near z_0, away from $\Sigma_1 \cap \Sigma_2$, P is of real principal type. The propagation of singularities for this case is well known, see [4].

If $O_k = \{t \in O : (-1)^k t > 0\}$, we define the four half-bicharacteristics at z_0 to be the curves

$$\gamma_{jk} : O_k \to \Sigma_j$$

defined by $\gamma_{jk} = \gamma_j|_{O_k}$, for $j, k = 1, 2$.

Our first main result is

THEOREM 1. *Let* P *satisfy* (1.1), (1.2), (1.3) *and let* $u \in \mathcal{D}'(X)$ *with* $z_0 \notin WF(Pu)$. *Suppose that for each* $j \in \{1,2\}$, *there exists* $k \in \{1,2\}$ *such that* $\gamma_{jk}(O_k) \cap WF(u) = \emptyset$. *Then* $z_0 \notin WF(u)$.

By assuming that S_P, the subprincipal symbol of P, avoids certain discrete sets of values, we can obtain more results. For each $j \in \{1,2\}$, we introduce the following conditions at z_0:

$$(1.4)_j \qquad \frac{(-1)^{j+1} i S_P(z_0)}{\{p_1, p_2\}(z_0)} - \frac{1}{2} \notin \{0, 1, 2, \cdots\} .$$

THEOREM 2. *Let* P *satisfy* (1.1), (1.2), (1.3), $(1.4)_j$ *and let* $u \in \mathcal{D}'(X)$ *with* $z_0 \notin WF(Pu)$. *Suppose that* $(\gamma_j(O) \setminus z_0) \cap WF(u) = \emptyset$. *Then* $z_0 \notin WF(u)$.

The conjunction of Theorems 1 and 2 yields the following:

COROLLARY. *Let* P *satisfy* (1.1), (1.2), (1.3). *Assume further that*

$$(1.5) \qquad \frac{i S_P(z_0)}{\{p_1, p_2\}(z_0)} - \frac{1}{2} \notin \{0, \pm 1, \pm 2, \cdots\} .$$

Then if Pu *is smooth at* z_0, *with* u *smooth along any two half-bicharacteristics at* z_0, *then* u *is smooth at* z_0.

REMARK 1. Let $k \in \{0, 1, 2, \cdots\}$. Then note that

$$(t\partial_t - k)(t^k \otimes \phi(x)) = 0 = (t\partial_t + k + 1)(\delta^{(k)}(t) \otimes \phi(x)) ,$$

for all $\phi \in \mathcal{D}'(\mathbf{R}^n)$. Hence the need for $(1.4)_j$ in Theorem 2.

REMARK 2. Our results are applicable to certain types of degenerate hyperbolic operators. If (x, t, ξ, τ) are coordinates in $T^*(\mathbf{R}^n \times \mathbf{R}) \setminus 0$, then examples of such operators are those with principal symbol $\tau^2 - t^2|\xi|^2$ (studied by Alinhac [2]), or $tp_2(x,t,\xi,\tau)$, where p_2 is strictly hyperbolic with respect to the hyperplanes $\{t = \text{constant}\}$ (studied by Hanges [7]).

2. *Reduction to canonical form*

Assumptions (1.1), (1.2), and (1.3) imply that after multiplying P by an operator elliptic near z_0, we can assume that the principal symbol of our operator has the form $r_0 r_1$, where r_j is real and positively homogeneous of degree j. Furthermore, we can assume that $\{r_1, r_0\} = 1$ in a conic neighborhood Γ of z_0. (See for example the complex noninvolutive case, Duistermaat-Sjöstrand [5].) Then by the homogeneous Darboux theorem, there exists an open cone $\tilde{\Gamma} \subset T^*(R^{n+1})\backslash 0$ and a homogeneous canonical transformation $\chi : \Gamma \to \tilde{\Gamma}$ such that $\chi(z_0) = (x_0, 0, \xi_0, 0) \in \tilde{\Gamma}$, (coordinates in $T^*(R^{n+1})\backslash 0$ are denoted by (x, t, ξ, τ)), and

$$(2.1) \qquad r_0 = t \circ \chi, \qquad r_1 = \tau \circ \chi .$$

Hence, after conjugation with Fourier integral operators associated to χ, we can reduce the study of P near z_0 to that of

$$(2.2) \qquad \tilde{L} = t\partial_t - \tilde{B}(x, t, D_x, D_t)$$

near $(x_0, 0, \xi_0, 0)$, where $\tilde{B} \in L_c^0(R^{n+1})$.

Finally we can reduce the operator $\tilde{L}$, near $(x_0, 0, \xi_0, 0)$, to the form

$$(2.3) \qquad L = t\partial_t - B(x, D_x)$$

where $B \in L_c^0(R^n)$ is properly supported. This can be done by conjugating L with pseudo-differential operators elliptic near $(x_0, 0, \xi_0, 0)$. This is essentially Proposition 2.1 of Hanges [7], and we refer to that paper for a proof. Note that this lemma has been obtained independently by Ivrii [11] and Melrose [15]. In the real analytic case this reduction has been obtained by Kashiwara, Kawai and Oshima [19].

Observe that if the principal symbol of B (resp. $\tilde{B}$) is b_0 (resp. $\tilde{b}_0$), then $b_0(x,\xi) = \tilde{b}_0(x,0,\xi,0)$.

Our problem has now been reduced to studying the propagation of singularities for L. We will do this by constructing explicit microlocal parametrices for L with chosen directions of propagation.

3. *A simple example*

To motivate the discussion in section 4 let us consider the simplest form the operator L can have. If $\lambda \in \mathbf{C}$, consider the operator $M_\lambda = t\partial_t - \lambda$. For $f \in C_0^\infty$, a formal solution to

$$M_\lambda u(x,t) = f(x,t), \qquad (x,t) \in \mathbf{R}^n \times \mathbf{R} \tag{3.1}$$

is given by

$$u(x,t) = t^\lambda \int^t s^{-\lambda-1} f(x,s)\, ds . \tag{3.2}$$

This formula can be made precise by introducing the limits from the upper half space

$$F_\lambda^+(t,s) = (t+i0)^\lambda \otimes (s+i0)^{-\lambda-1} . \tag{3.3}$$

We then define the kernel

$$K_\lambda(x,t,y,s) = [F_\lambda^+(t,s) \cdot H(t-s)] \otimes \delta(x-y) , \tag{3.4}$$

where H is the Heaviside function. The multiplication is well defined by noting that the wave front of $(t+i0)^\lambda$ is at most the top half of the fiber over the origin. So K_λ is an exact two sided inverse for M_λ. Moreover, it is an easy exercise to see that $WF'(K_\lambda) \subset \mathcal{C}_1$, ($\mathcal{C}_1$ is defined in section 4). Geometrically speaking, this means that K_λ propagates singularities along the bicharacteristic strips of the functions t and τ in certain directions. Actually, four inverses exist by also considering the limits from the lower half space,

$$[F_\lambda^\pm(t,s) \cdot H(t \pm s)] \otimes \delta(x-y) \tag{3.5}$$

and the singularities can be estimated as above. So we have one two-sided parametrix for each possible choice of direction along the bicharacteristic strips of t and τ. Now the application of a standard argument (see Duistermaat-Hörmander [4] or Hanges [8]) proves Theorem 1 for the operator M_λ.

If we are willing to impose conditions on λ, then a solution to (3.1) is given by

$$u(x,t) = \int_0^1 \rho^{-\lambda-1} f(x,\rho t)\, d\rho \tag{3.6}$$

provided that $\operatorname{Re}\lambda < 0$. Note that here u is smooth if f is, and that the kernel corresponding to this formula propagates singularities quite differently from K_λ. This fact motivates the discussion in section 5.

4. *Results independent of the lower order terms*

Previous considerations allow us to reduce the study of P near z_0, to the study of $L = t\partial_t - B(x, D_x)$ near $(x_0, 0, \xi_0, 0)$. The main result of this section is Proposition 4.1, which implies Theorem 1 by the invariance of the bicharacteristic flow under homogeneous canonical transformations.

Let $(x_0, \xi_0) \in T^*(R^n)\backslash 0$. Then we define for $i, j = 1, 2$

$$\mathcal{L}_{ij}(x_0,\xi_0) = \{(x_0, t, \xi_0, \tau) \in T^*(R^{n+1})\backslash 0 : t\tau = 0 \text{ and } ((-1)^i t > 0 \text{ or } (-1)^j \tau > 0)\}. \tag{4.1}$$

Proposition 4.1. *Let* $u \in \mathcal{D}'(R^{n+1})$ *with* $(x_0, 0, \xi_0, 0) \notin WF(Lu)$. *If there exists* (i, j) *such that* $\mathcal{L}_{ij}(x_0, \xi_0) \cap WF(u) = \emptyset$, *then* $(x_0, 0, \xi_0, 0) \notin WF(u)$.

Proposition 4.1 will follow after constructing four left parametrices for L (one for each pair (i, j)). The first step is the introduction of a class of symbols.

Let $G_+ = C\backslash\{z : z = iy,\ y \le 0\}$ and let $\mathcal{H}(G_+ \times G_+;\ S^m_{1,0}(R^n \times R^n))$ be the space of holomorphic functions on $G_+ \times G_+$ with values in $S^m_{1,0}(R^n \times R^n)$, equipped with the obvious semi-norms.

Next we define $\mathcal{B}^N(G_+ \times G_+;\, S^m_{1,0}(R^n \times R^n))$ to be the subspace of all $a \in \mathcal{H}(G_+ \times G_+;\, S^m_{1,0}(R^n \times R^n))$ such that for all non-negative integers j, k, multi-indices α, β, compact $K \subset\subset R^n$ and bounded $\Omega \subset G_+ \times G_+$ there exists $C(j, k, \alpha, \beta, K, \Omega) > 0$ such that

$$(4.2) \qquad |\partial_z^j \partial_w^k \partial_x^\alpha \partial_\theta^\beta a(z, w, x, \theta)| \le \frac{C(1+|\theta|)^{m-|\beta|}}{|z|^{N+j}\,|w|^{N+k}}$$

for all $\theta \in R^n$, $x \in K$, $(z, w) \in \Omega$. $\mathcal{B}^N$ is a Fréchet subspace of $\mathcal{H}$ taking as additional semi-norms the smallest constants making (4.2) true. The symbol space $\mathcal{B}^N$ satisfies the usual symbolic calculus, (asymptotic sums, etc.), see Hanges [8].

EXAMPLE. If $|b_0(x, \theta)| < C_0$ for all $(x, \theta) \in T^*(R^n)\backslash 0$ (which we shall assume from now on), then we have

$$w^{b_0(x,\theta)}\, z^{-b_0(x,\theta)-1} \in \mathcal{B}^{C_0+1+\alpha}(G_+ \times G_+;\, S^0_{1,0}(R^n \times R^n)), \text{ for all } \alpha > 0 .$$

Now we are ready to construct distribution kernels that will be used to compute parametrices for L. For $a \in \mathcal{B}^N(G_+ \times G_+;\, S^m_{1,0}(R^n \times R^n))$ we define $E_a^+ \in \mathcal{D}'((R^n \times R) \times (R^n \times R))$ by

$$(4.3) \qquad E_a^+(x, t, y, s) = \lim_{\substack{\varepsilon \to 0^+ \\ \eta \to 0^+}} \int_{R^n} e^{i<x-y,\theta>}\, a(s+i\varepsilon, t+i\eta, x, \theta)\, d\theta \ .$$

For each $\varepsilon > 0$, $\eta > 0$, (4.3) is a usual oscillatory integral. The limit is taken in $\mathcal{D}'$ and exists by the definition of $\mathcal{B}^N$. We also write

$$E_a^+(x, t, y, s) = \int_{R^n} e^{i<x-y,\theta>}\, a(s+i0, t+i0, x, \theta)\, d\theta \ .$$

Note that away from $\{s=0\} \cup \{t=0\}$, E_a^+ is a Fourier integral distribution.

A parametrix for L will have the form $H(t-s)\cdot E_a^+(x,t,y,s)$, where H is the Heaviside function. This multiplication is well defined by virtue of the multiplication formula of Hörmander [9], and we have

PROPOSITION 4.2. $WF'(H(t-s)\cdot E_a^+) \subset \mathcal{C}_1 \cup \mathcal{C}_2 \subset T^*(\mathbf{R}^{2n+2})\backslash 0$ *where we define*

$$\begin{aligned}\mathcal{C}_1 &= \{(x,t,\xi,\tau;\ x,s,\xi,\sigma): t \geq s,\ t\tau = s\sigma = 0,\ \tau \geq 0,\ \sigma \leq 0\} \\ &\cup \{(x,0,\xi,\tau;\ x,0,\xi,\sigma): \tau \geq \sigma\} \cup \textit{Diagonal},\end{aligned}$$

$$\begin{aligned}\mathcal{C}_2 &= \{(x,t,0,\tau;\ y,s,0,\sigma): t \geq s,\ t\tau = s\sigma = 0,\ \tau \geq 0,\ \sigma \leq 0\} \\ &\cup \{(x,0,0,\tau;\ y,0,0,\sigma): \tau \geq \sigma\} \cup \{(x,t,0,\tau;\ y,t,0,\tau): \tau \neq 0\}.\end{aligned}$$

Note that since we are working near $(x_0, 0, \xi_0, 0)$, $\mathcal{C}_1$ is the set of interest to us. ($\mathcal{C}_2$ plays virtually no role in our study.) Geometrically, this means that singularities are propagated along the bicharacteristic strips of t and τ in given directions.

Now by applying a modified version of the method of geometrical optics, we can obtain a right parametrix for L of the form $H(t-s)\cdot E_a^+$. We state this as

PROPOSITION 4.3. *For every* $a > 0$ *there exists* $a \in \mathcal{B}^{C_0+1+a}(G_+ \times G_+;\ S^0_{1,0}(\mathbf{R}^n \times \mathbf{R}^n))$ *such that*

$$L(H(t-s)\cdot E_a^+) = I + R$$

with $WF'(R) \subset \mathcal{C}_2$.

The methods discussed above show that we actually have four right parametrices for L. By introducing the limits from the lower half space

$$E_a^-(x,t,y,s) = \int_{\mathbf{R}^n} e^{i\langle x-y,\theta\rangle} a(s-i0,\ t-i0,\ x,\ \theta)\, d\theta$$

we can define inverses for L by $H(t \pm s) E_a^{\pm}$ for suitable choices of a. Each of these four kernels has a different direction of propagation which is needed to prove Proposition 4.1. Left parametrices can be computed by transposition, or directly using a slightly different symbol class. The existence of these left parametrices implies Proposition 4.1, hence Theorem 1.

5. *Results depending on the lower order terms*

In this section we construct microlocal parametrices for L assuming that the subprincipal symbol avoids certain discrete sets of values. The fact that this phenomenon can affect the regularity of solutions has been observed for a long time. See for example Alinhac [1], [2], Baouendi-Goulaouic [3], Treves [18] and references cited there. Parametrices similar to those constructed here were first used by Hanges [6], [7] in studying solvability for a class of degenerate hyperbolic operators. The main difference here is that left parametrices are needed. The constructions of this section are essentially the same as those in [6], [7]. The main result of this section is Proposition 5.1 which implies Theorem 2.

PROPOSITION 5.1. *Let* $u \in \mathcal{D}'(R^{n+1})$ *with* $(x_0, 0, \xi_0, 0) \notin WF(Lu)$. *If we have* $b_0(x_0, \xi_0) \notin \{0, 1, 2, \cdots\}$ *and* $\{(x_0, 0, \xi_0, \tau) : \tau \neq 0\} \cap WF(u) = \emptyset$, *then* $(x_0, 0, \xi_0, 0) \notin WF(u)$.

We begin by recalling the symbol class $\tilde{S}^m(I \times R^n \times R^n) \subset C^\infty(I \times R^n \times R^n)$ where $I =]0, 1[\subset R$ and $m \in R$, (see [7]). We say $a(\rho, y, \theta) \in \tilde{S}^m$ if and only if for all $\varepsilon > 0$, non-negative integers j, multi-indices α, β and compact $K \subset R^n$ we have

$$\sup_{\substack{\rho \in I \\ y \in K \\ \theta \in R^n}} \left| \frac{\rho^{\varepsilon} (\rho \partial_\rho)^j (\partial_y)^\alpha (\partial_\theta)^\beta a(\rho, y, \theta)}{(1 + |\theta|)^{m - |\beta|}} \right| < \infty . \tag{5.1}$$

$\tilde{S}^m$ is a Frechet space with semi-norms determined by (5.1). Again a usual symbolic calculus holds.

If we assume $\operatorname{Re} b_0(x,\xi) < 0$ for all $(x,\xi) \in T^*(R^n)\backslash 0$, then for each $a \in \tilde{S}^m$ we define $T_a \in \mathcal{D}'(R^{2n+2})$ as follows:

(5.2) if $\phi(x,t,y,s) \in C_0^\infty(R^n \times R \times R^n \times R)$, then $<T_a, \phi> =$

$$\int_0^1 \int_{R^{3n+1}} e^{i<x-y,\theta>} \rho^{-b_0(y,-\theta)-1} a(\rho,y,\theta)\phi(x,t,y,\rho t) \frac{d\theta\, dx\, dt\, dy\, d\rho}{(2\pi)^n}.$$

This makes sense as an oscillatory integral by making partial integrations with respect to x. Note that $T_a : C_0^\infty(R^{n+1}) \to C^\infty(R^{n+1})$. These operators will be used to construct left parametrices for L, hence we need information on $WF(T_a)$.

LEMMA 5.2. *$WF'(T_a) \subset \mathcal{G} \cup \mathcal{F} \subset T^*(R^{2n+2})\backslash 0$ where*

$$\mathcal{G} = \{(x,t,\xi,\tau;\ x,s,\xi,\sigma) : t\tau = s\sigma = 0,\ \tfrac{\tau}{\sigma} \in [0,1],\ \tfrac{s}{t} \in [0,1]\}$$
$$\cup\ \textit{Diagonal},$$

$$\mathcal{F} = \{(x,t,0,\tau;\ y,0,0,\sigma) : t\tau = 0,\ \tfrac{\tau}{\sigma} \in [0,1]\}$$
$$\cup\ \{(x,t,0,\tau;\ y,t,0,\tau) : \tau \neq 0\}.$$

We say $\frac{\tau}{0} \in [0,1]$ if and only if $\tau = 0$.

Lemma 5.2 tells us that T_a propagates singularities quite differently from the kernels studied in section 4. Remember $\mathcal{G}$ is the important set; $\mathcal{F}$ plays no role. The next lemma gives a sufficient condition for which L will have a left parametrix of the form T_a.

LEMMA 5.3. *Assume that $\operatorname{Re} b_0(x,\xi) < 0$ for every $(x,\xi) \in T^*(R^n)\backslash 0$. Then there exists $a \in \tilde{S}^0$ such that*

$$WF'(T_a L - I) \subset \mathcal{F}.$$

Again the proof is based on the method of geometrical optics, see [8].

We are now ready to construct a left parametrix for L which will imply Proposition 5.1.

If we assume $b_0(x_0, \xi_0) \notin \{0, 1, 2, \cdots\}$, then there exists an open cone $V \subset T^*(R^n)\backslash 0$ such that $(x_0, \xi_0) \in V$ and $b_0(x, \xi) \notin \{0, 1, 2, \cdots\}$ for all $(x, \xi) \in \bar{V}$.

PROPOSITION 5.4. *Let* $b_0(x_0, \xi_0) \notin \{0, 1, 2, \cdots\}$. *Then there exist distribution kernels* K *and* R *such that*

$$KL = I + R$$

where $WF'(K) \subset \mathcal{G} \cup \mathcal{F}$ *and*

$$WF'(R) \subset \mathcal{F} \cup \{(x, t, \xi, 0;\ x, 0, \xi, \sigma) : (x, \xi) \notin \bar{V}\}.$$

Proof. Since we assume b_0 bounded, we can choose a positive integer ℓ so that $\mathrm{Re}\, b_0(x, \xi) - \ell < 0$ for all $(x, \xi) \in T^*(R^n)\backslash 0$. So by Lemma 5.3 we have $a \in \tilde{S}^0$ such that

$$T_a(t\partial_t - (B - \ell)) = I + R$$

with $WF'(R) \subset \mathcal{F}$.

The fact that $b_0(x_0, \xi_0) \notin \{0, 1, 2, \cdots\}$ implies the existence of $(j - B)^{-1} \in L^0_c(R^n)$ such that $(j - B)^{-1}(j - B) = I + S_j$ and $WF(S_j) \cap \bar{V} = \emptyset$.

Next we introduce the jth trace γ_j and the ℓth order Taylor remainder M_ℓ. The operators are defined by the following identity: if u is sufficiently smooth we have

$$u = \sum_{j=0}^{\ell-1} \frac{t^j}{j!} \gamma_j u + t^\ell M_\ell \partial_t^\ell u. \tag{5.3}$$

We define K to be

$$K = \sum_{j=0}^{\ell-1} \frac{t^j}{j!} (j - B)^{-1} \gamma_j + t^\ell M_\ell T_a \partial_t^\ell. \tag{5.4}$$

Note that $WF'(M_\ell) \subset \mathcal{G} \cup \mathcal{F}$, see [7]. Finally we have

$$\text{(5.5)} \qquad KL = I + \sum_{j=0}^{\ell-1} \frac{t^j}{j!} (S_j \otimes I) \circ \gamma_j + t^\ell M_\ell R \partial_t^\ell$$

with

$$WF'((S_j \otimes I) \circ \gamma_j) \subset \mathcal{F} \cup \{(x, t, \xi, 0; x, 0, \xi, \sigma) : (x, \xi) \notin \bar{V}\},$$

$$WF'(M_\ell R) \subset \mathcal{F}. \qquad \text{q.e.d.}$$

Note that Ku is defined for $u \in \mathcal{E}'(\mathbf{R}^{n+1})$ provided that WF(u) is sufficiently close to $(x_0, 0, \xi_0, 0)$. Hence Proposition 5.1 follows from Proposition 5.4. Theorem 2 follows by the invariance of S_P at the double characteristics.

REMARK. We point out here that if we assume the operators $B - j$ are all hypoelliptic at (x_0, ξ_0), then a minor modification of the proof given above shows that Proposition 5.1 is still true, although the parametrix (5.4) need not exist.

In fact, hypoellipticity is the correct notion here. For assume that there exists k such that $k - B$ is not hypoelliptic at (x_0, ξ_0). Then choose $u(x) \in \mathcal{D}'(\mathbf{R}^n)$ such that $(x_0, \xi_0) \in WF(u)$ but $(x_0, \xi_0) \notin WF((k-B)u)$. If we define $v = t^k \otimes u(x)$, we have $Lv = t^k \otimes (k - B)u$. Hence Lv is smooth at $(x_0, 0, \xi_0, 0)$, $WF(v) \cap \{(x_0, 0, \xi_0, \tau) : \tau \neq 0\} = \emptyset$, but $(x_0, 0, \xi_0, 0) \in WF(v)$.

REMARK. By employing the right parametrices of section 4, it is easy to show (under the correct assumption on the lower order terms) that there always exists $u \in \mathcal{D}'(\mathbf{R}^{n+1})$ such that Lu is smooth at $(x_0, 0, \xi_0, 0)$ and WF(u) is equal to three half-bicharacteristics at $(x_0, 0, \xi_0, 0)$.

REFERENCES

[0] S. Alinhac, "Parametrix pour un systeme hyperbolique à multiplicité variable," Comm. P.D.E., 2(3), 251-296 (1977).

[1] ———, "Problèmes de Cauchy pour des opérateurs singuliers," Bull. Soc. Mat. France, 102 (1974), 289-315.

[2] S. Alinhac, "Parametrix et propagation des singularités pour un problème de Cauchy a multiplicité variable," Astérisque, 34-35 (1976), 3-26.

[3] M. S. Baouendi and C. Goulaouic, "Cauchy problems with characteristic initial hypersurface," Comm. Pure and Appl. Math., 26 (1973), 455-475.

[4] J. J. Duistermaat and L. Hörmander, "Fourier Integral operators II," Acta. Math., 128 (1972), 183-269.

[5] J. J. Duistermaat and J. Sjöstrand, "A Global construction for pseudo-differential operators with non-involutive characteristics," Inventiones Math., 20 (1973), 209-225.

[6] N. Hanges, Thesis, Purdue University, December, 1976.

[7] ———, "Parametrices and local solvability for a class of singular hyperbolic operators," Comm. P.D.E., 3(2), 105-152 (1978).

[8] ———, "Parametrices and propagation of singularities for operators with non-involutive characteristics," to appear.

[9] L. Hörmander, "Fourier Integral operators I," Acta. Math., 127 (1971), 79-183.

[10] ———, "Pseudodifferential operators and hypoelliptic equations," Proceedings of symposia in Pure Mathematics, X (1967), 138-183.

[11] V. Ya. Ivrii, "Wave fronts of solutions of certain pseudodifferential equations," Functional Analysis and its applications, 10 (1976), 141-142.

[12] ———, "Wave fronts of solutions to some microlocally hyperbolic pseudo-differential equations," Soviet Math. Dokl. 17 (1976), 233-236.

[13] ———, "Wave fronts of solutions of symmetric pseudodifferential systems," Soviet Math. Dokl. 18 (1977), 540-544.

[14] R. Melrose, "Normal self-intersection of the characteristic variety," Bull. Amer. Math. Soc., 81 (1975), 939-940.

[15] ———, "Non-involutive self-intersections of the characteristic variety," M.I.T. preprint.

[16] ———, "Equivalence of glancing hypersurfaces," Inventiones Math. 37 (1976), 165-191.

[17] T. Miwa, "Propagation of micro-analyticity for solutions of pseudo-differential equations, I," Publ. RIMS, Kyoto Univ. 10 (1975), 521-533.

[18] F. Treves, "Second order Fuchsian elliptic equations and eigenvalue asymptotics," Lecture Notes in Math. # 459, 283-340.

[19] M. Kashiwara, T. Kawai and T. Oshima, "Structure of cohomology groups whose coefficients are microfunction solution sheaves of systems of pseudo-differential equations with multiple characteristics, I and II," Proc. Japan Acad. 50 (1974), 420-425 and 549-550.

[20] M. Kashiwara, T. Kawai and T. Oshima, "Structure of a single pseudo-differential equation in a real domain," J. Math. Soc. Japan, 28 (1) 1976, 80-85.

SUBELLIPTIC OPERATORS

Lars Hörmander[1)]

1. *Introduction*

Let Ω be an open set in $\mathbf{R}^n$ (or a manifold of dimension n) and let P be a pseudo-differential operator of order m in Ω with principal symbol p. It is well known that P is elliptic if and only if

$$u \in \mathcal{D}'(\Omega),\ Pu \in H^{loc}_{(s)}(\Omega) \implies u \in H^{loc}_{(s+m)}(\Omega)\,. \tag{1.1}$$

Here $H^{loc}_{(s)}(\Omega)$ is the space of distributions f in Ω such that the Fourier transform of ϕf is in L^2 with respect to the measure $(1+|\xi|^2)^s d\xi$ if $\phi \in C_0^\infty(\Omega)$. Since $u \in H^{loc}_{(s+m)}(\Omega)$ implies $Pu \in H^{loc}_{(s)}(\Omega)$ it is clear that (1.1) is the best possible regularity result.

One calls P subelliptic with loss of δ derivatives if $0 < \delta < 1$ and

$$u \in \mathcal{D}'(\Omega),\ Pu \in H^{loc}_{(s)}(\Omega) \implies u \in H^{loc}_{(s+m-\delta)}(\Omega)\,. \tag{1.2}$$

The reason for the condition $\delta < 1$ is that (1.2) is then independent of the lower order terms in P. In fact, assume that (1.2) is valid and that Q is of order $m-1$. If $(P+Q)u \in H^{loc}_{(s)}$ and $u \in H^{loc}_{(t)}$ it follows that $Pu \in H^{loc}_{(s')}$ for $s' = \min(s, t-m+1)$. Hence $u \in H^{loc}_{(t')}$ where $t' = \min(s+m-\delta, t+1-\delta)$. A finite number of iterations of the argument give $u \in H^{loc}_{(s+m-\delta)}$ since $\delta < 1$.

---

1) Supported in part by NSF contract MCS77-18723.

Seminar on Singularities of Solutions
0-691-08221-9/78/0127 $04.10/1 (cloth)
0-691-08213-8/78/0127 $04.10/1 (paperback)
For copying information, see copyright page

It was proved in [4] that (1.2) implies ellipticity if $\delta < 1/2$ and that (1.2) is valid with $\delta = 1/2$ if and only if

$$\{\mathrm{Re}\, p, \mathrm{Im}\, p\} > 0 \quad \text{when} \quad p = 0\ . \tag{1.3}$$

On the other hand, if $\{\mathrm{Re}\, p, \mathrm{Im}\, p\} < 0$ for some zero of p, then (1.2) is not valid for any δ. Egorov [1] announced an extension of these results: The set of subelliptic operators with loss of δ derivatives is constant for $k/(k+1) \leqq \delta < (k+1)/(k+2)$, if k is a positive integer, and it is characterized by conditions generalizing (1.3) which involve repeated Poisson brackets of at most $k+1$ factors $\mathrm{Re}\, p$ and $\mathrm{Im}\, p$. (See Theorem 3.4 for the precise conditions.) His proofs were published in [2, 3]. One part of the necessary conditions was found independently by Nirenberg and Treves [9], and the sufficiency was also proved by Treves [10] in the case of differential operators. The purpose of this paper is to give a presentation of Egorov's arguments with modifications which allow us to avoid some delicate and seemingly obscure points. These changes will be discussed in section 8.

In section 2 we shall study the Taylor expansion of the principal symbol of a subelliptic operator. This prepares for the proof of necessary conditions in section 3 and serves as a simple model for the discussion of the local properties in section 4. There we also define a partition of unity which is used to localize the estimates, which are proved in section 6 after a study of the localized operators in section 5. In section 7 we have collected calculus lemmas which are important in the study of the properties of the principal symbol.

The work leading to this paper was begun in collaboration with Anders Melin and I owe to him several arguments used in section 4 as well as the stimulation of many discussions of Egorov's work. I would also like to thank him and H. M. Maire for their careful study of a preliminary manuscript of this paper.

2. *The Taylor expansion of the principal symbol*

As recalled in the introduction subellipticity (or even hypoellipticity) of P requires that the principal symbol $p = p_1 + ip_2$ satisfies the condition

$$\{p_1, p_2\} \geqq 0 \text{ when } p_1 = p_2 = 0 . \tag{2.1}$$

(See also [7, Theorem 7.3].) Here $\{p_1, p_2\}$ is the Poisson bracket,

$$\{p_1, p_2\} = \sum (\partial p_1/\partial\xi_j \, \partial p_2/\partial x_j - \partial p_1/\partial x_j \, \partial p_2/\partial \xi_j) .$$

This condition is obviously invariant under canonical transformations and multiplication of p by a non-vanishing factor. Other necessary conditions will be introduced later on when their invariance becomes evident.

If f is a function we shall denote its Hamilton vector field by H_f, thus $H_f g = \{f, g\}$. We shall use the notation $I = (i_1, \cdots, i_j)$ to denote a sequence of $j = |I|$ elements each of which is 1 or 2. We shall write $H_i = H_{p_i}$ and

$$H^I = H_{i_1} \cdots H_{i_j}, \; (\mathrm{ad}\, H)^I = \mathrm{ad}\, H_{i_1} \cdots \mathrm{ad}\, H_{i_j}, \; p_I = H_{i_1} \cdots H_{i_{j-1}} p_{i_j}$$

where for vector fields v, w the operator $\mathrm{ad}\, v$ is defined by

$$(\mathrm{ad}\, v) w = [v, w] .$$

By the Jacobi identity

$$H_{\{f,g\}} = [H_f, H_g] = \mathrm{ad}\, H_f \, H_g$$

and since $\{f, g\} = H_f g$, repeated use of this identity shows that the Hamilton field of $H^I p_2$ is $(\mathrm{ad}\, H)^I H_2$.

Our goal is to study the Taylor expansion of p at a point γ. No homogeneity condition will be used so p may be thought of as a complex valued function on any symplectic manifold. Recall that $\{p_1, p_2\} = \sigma(H_1, H_2)$

where σ is the symplectic form; with the standard coordinates in a cotangent bundle we have $\sigma = \sum d\xi_j \wedge dx_j$. The following integers will play an important role:

(2.2) $k = \sup\{j \in \mathbf{Z};\ p_I(\gamma) = 0 \text{ when } |I| \leqq j\}$

(2.3) $s = \sup\{j \in \mathbf{Z};\ j \leqq k$ and all commutators of H_1 and H_2 with at most j factors are linearly dependent at $\gamma\}$.

Here H_1 and H_2 are regarded as commutators with 1 factor. Thus $k = 0$ means that $p \neq 0$ while $k = 1$ means that $p = 0$ but $\{p_1, p_2\} \neq 0$. If $k \neq 0$ then $s = 0$ means that H_1 and H_2 are linearly independent. The definitions are obviously invariant under canonical transformations. Moreover, if we multiply p by a non-vanishing complex valued factor a, then the Poisson brackets of $\operatorname{Re} ap$ and $\operatorname{Im} ap$ of order $\leqq j$ are linear combinations with smooth coefficients of those of p_1 and p_2, and vice versa. This shows that the numbers k and s are also invariant under multiplication by non-vanishing factors.

When $k = 0$ we can reduce p to the constant 1 by multiplication with p^{-1}. If $k = 1$ then $s = 0$ since $\sigma(H_1, H_2) \neq 0$. After multiplication by a non-vanishing function one can then introduce symplectic coordinates vanishing at γ so that $p(x,\xi) = \xi_1 + ix_1$ (see [7, Theorem 7.1]). We shall therefore assume $k > 1$ in what follows.

If $H_1 = H_2 = 0$ at γ then $k = s = \infty$ and there is not much more to say in that case. Thus we assume that for example $H_1 \neq 0$. Then p_1 can be taken as the symplectic coordinate ξ_1 so $p(x,\xi) = \xi_1 + ip_2(x,\xi)$. By the Malgrange preparation theorem we can write near 0

$$\xi_1 = a(x,\xi)(\xi_1 + ip_2(x,\xi)) - r_1(x,\xi') - ir_2(x,\xi')$$

where $\xi' = (\xi_2, \cdots, \xi_n)$ and $a(0) \neq 0$. We can take $\eta_1 = \xi_1 + r_1(x,\xi')$ and $y_1 = x_1$ as new symplectic coordinates since their Poisson bracket is 1. Now $0 = \partial r_2/\partial\xi_1 = \{r_2, x_1\} = \{r_2, y_1\}$ which means that r_2 becomes independent of η_1 in the new variables. Changing notations we may therefore always assume that

$$p_1(x,\xi) = \xi_1,\ p_2(x,\xi) \text{ is independent of } \xi_1 .$$

To keep track of all the terms which arise when one forms Poisson brackets of high order it is convenient to introduce weights of polynomials in symplectic coordinates $x_1, \xi_1, \cdots, \xi_n$. If positive weights $m_1, \mu_1, \cdots \mu_n$ are attached to these variables we shall say that a C^∞ function F vanishes of weight k at 0 if

$$D_\xi^\alpha D_x^\beta F(0) = 0 \text{ when } \langle\alpha,\mu\rangle + \langle\beta, m\rangle < k .$$

It is clear that a product F_1F_2 vanishes of weight $k_1 + k_2$ if F_1 and F_2 vanish of weight k_1 and k_2. If F vanishes of weight k then $D_\xi^\alpha D_x^\beta F$ vanishes of weight $k - \langle\alpha,\mu\rangle - \langle\beta, m\rangle$.

LEMMA 2.1. *Suppose that one can find symplectic coordinates* x, ξ *with the origin at* y *and positive weights* m_j, μ_j *with* $m_j + \mu_j = k+1$ *for* $j = 1, \cdots, n$ *such that* p_1 *and* p_2 *vanish of weight* k *at* 0. *Then it follows that* $H^I p_2$ *vanishes of weight* $k - |I|$ *at* 0 *if* $k < |I|$, *and that* $H^I p_2(0)$ *only depends on the derivatives of* p_1 *and* p_2 *of weight* k *at* 0 *if* $|I| = k$.

Proof. From the definition of Poisson brackets

$$H_f g = \{f, g\} = \sum (\partial f/\partial \xi_j\, \partial g/\partial x_j - \partial f/\partial x_j\, \partial g/\partial \xi_j)$$

it follows that if f and g vanish of weights w_f and w_g respectively at 0, then each term vanishes of weight $w_f + w_g - m_j - \mu_j = w_f + w_g - k - 1$. If $w_f = k$ then this weight is $w_g - 1$, so the lemma follows immediately by induction for increasing $|I|$.

The definition (2.2) of k implies that

$$(2.2)' \qquad (\partial/\partial x_1)^j\, p_2(0) = 0 \text{ if } j < k .$$

We can therefore apply Lemma 7.2 to p_2 with $t = x_1$ and $y = (x', \xi')$. Two cases occur depending on whether the linear form L in the lemma is 0 or not. We first consider the simple case where $L = 0$. (Lemma 7.2 will not really be used at first but it is the reason for the appearance of two cases.)

THEOREM 2.2. *Let* p_1 *and* p_2 *vanish at* γ *and assume that with* k *and* s *defined by* (2.2) *and* (2.3) *we have* $0 < k \leqq 2s < \infty$. *For arbitrary functions* $f_1, \cdots, g_2 \in C^\infty$ *we have then*

$$(2.4)\quad (H_{f_1p_1+f_2p_2})^k(g_1p_1+g_2p_2)(\gamma) = c(f_1g_2-f_2g_1)(f_1a_1+f_2a_2)^{k-1}(\gamma)\,.$$

Here $0 \neq c \in \mathbf{R}$, $0 \neq (a_1, a_2) \in \mathbf{R}^2$ *and* $a_2H_1 - a_1H_2 = 0$ *at* γ.

Proof. The statement is obviously invariant under canonical transformations and multiplication of $p_1 + ip_2$ by a non-vanishing factor. We may therefore assume that $p_1 = \xi_1$ and that p_2 is independent of ξ_1. From (2.2) and (2.3) we obtain

$$(2.5)\qquad (\partial/\partial x_1)^j p_2(0) = 0 \ \text{ if } \ j < k;\ d(\partial/\partial x_1)^j p_2(0) = 0 \ \text{ if } \ j < s\,.$$

(Note that $j < k/2$ implies $j < s$.) It follows that p_1 and p_2 vanish at 0 of weight k if the weights of the coordinates are chosen to be

$$(2.6)\qquad m_1 = 1,\ \mu_1 = k,\ m_j = \mu_j = (k+1)/2 \ \text{ if } \ j > 1\,.$$

The only term in p_2 of weight k is of the form $bx_1{}^k$ so the coefficient b here must be different from 0 since Lemma 2.1 would otherwise show that all Poisson brackets with k+1 factors are equal to 0. When computing the left hand side of (2.4) we may by Lemma 2.1 replace $f_1, \cdots, g_2$ by their values at 0 and p_2 by $bx_1{}^k$. Note that $k > 1$ since $s > 0$. It follows that the left hand side of (2.4) is equal to

$$(f_1g_2 - f_2g_1)\,f_1{}^{k-1}\,b\,k!$$

which proves Theorem 2.2.

When k is odd the sign of the coefficient c in (2.4) is clearly an invariant, and the second necessary condition for subellipticity is that $c > 0$ then, or more generally

$$(2.7) \qquad p_1 = 0,\ H_1{}^i p_2 = 0 \ \text{ for } \ i < j \ \text{ implies } \ j \ \text{ even or } \ H_1{}^j p_2 \geqq 0\ .$$

That (2.7) is necessary for hypoellipticity is well known (see e.g. [7, Theorem 7.3]). To relate (2.7) to Theorem 2.2 we first observe that if (2.7) is false then the restriction of p_2 to an integral curve of H_1 where $p_1 = 0$ changes sign from $+$ to $-$ at some zero γ of finite order. In any neighborhood of γ there is another zero γ' of p such that this sign change occurs with minimal multiplicity μ on the integral curve of H_1 through γ'. As in [7, Theorem 7.1] we conclude that the product of p and some non-vanishing factor can be written in the form

$$\xi_1 - i x_1{}^{\mu}$$

with appropriate symplectic coordinates. Here μ is odd, $\mu > 1$, by (2.1). It follows that the numbers k and s in (2.2) and (2.3) are μ and $\mu - 1$ respectively so we obtain the situation in Theorem 2.2 with $c < 0$. Thus (2.7) is equivalent to the condition that *the coefficient* c *in Theorem 2.2 is positive if* k *is odd.* In particular this shows that (2.7) is invariant under multiplication of $p_1 + ip_2$ by a non-vanishing factor.

REMARK. The proof of Theorem 2.2 showed that if $p_1 = \xi_1$ and $p_2 = p_2(x, \xi')$ then $k(0) \geqq m$ and $s(0) \geqq m/2$ if

$$(2.5)' \quad (\partial/\partial x_1)^j p_2(0) = 0 \ \text{ when } \ j < m \ \text{ and } \ d(\partial/\partial x_1)^j p_2(0) = 0 \ \text{ when } \ j < m/2\,.$$

Indeed, with the weights given by (2.6) with k replaced by m, the Poisson brackets with j factors vanish at 0 of weight $m - j + 1$, so they are 0 if $j \leqq m$. Moreover, apart from powers of x_1 they vanish of weight $m - j + 3/2$ at 0. Hence they contain no first order terms in (x', ξ') if $m - j + 3/2 > (m+1)/2$, that is, $j < m/2 + 1$. Thus $s \geqq m/2$ if m is even and $s \geqq (m+1)/2$ if m is odd, which can be summarized by $s \geqq m/2$.

We shall now pass to the more difficult case where Theorem 2.2 is not applicable, that is,

$$k > 2s . \tag{2.8}$$

As indicated before we exclude the elementary and exceptional case where $k = 1$ and $s = 0$. It follows from (2.2) and (2.3) that

(2.9) $(\partial/\partial x_1)^j p_2(0) = 0$ for $j < k$; $d(\partial/\partial x_1)^j p_2(0) = 0$ when $j < s$.

In addition

(2.9)′ $$d(\partial/\partial x_1)^s p_2(0) \neq 0 .$$

In fact, otherwise we could take $m = 2s+1$ in the preceding remark and conclude that the number defined by (2.3) is at least $s + 1/2$, which is a contradiction.

By a symplectic orthogonal transformation of the $x'\xi'$ variables we can attain that $d(\partial/\partial x_1)^s p_2(0) = d\xi_2$. If we apply Lemma 7.2 it follows that

$$p_2(x, \xi') = (x_1{}^s + 0(x_1{}^{s+1}))\xi_2/s! + \mathcal{O}(|x_1|^k + |x'|^2 + |\xi'|^2) \tag{2.10}$$

if k is finite; otherwise k should be replaced by any integer satisfying (2.8). To be able to apply Lemma 2.1 we would now like to choose weights so that ξ_1 and p_2 vanish of weight k at 0. This requires that

(2.6)′ $$\mu_1 = k,\ m_1 = 1,\ \mu_2 = k-s,\ m_2 = s+1,\ m_j = \mu_j = (k+1)/2 \text{ if } j > 2 .$$

Each choice motivates the following one. However, it does not follow from (2.10) that p_2 vanishes of weight k then, for m_2 may be small compared to k, so terms involving x_2 have to be examined more carefully by extracting more information from the definition (2.2) of s. To do so we make a more precise canonical transformation by introducing $\partial^s p_2(0, x', \xi')/\partial x_1{}^s$ as a new ξ_2 variable. Thus we may assume that

$$\partial^s p_2(0, x', \xi')/\partial x_1{}^s = \xi_2 . \tag{2.11}$$

Now (2.2) implies that

$$((\mathrm{ad}\, H_1)^s H_2)^j H_1{}^i p_2 = 0 \text{ at } 0 \text{ if } i+(s+1)j < k$$

and $(\mathrm{ad}\, H_1)^s H_2$ is the Hamilton field of $H_1{}^s p_2$. We may replace it by the Hamilton field of $H_1{}^s p_2(0, x', \xi')$ for both can be considered as vector fields in the plane $x_1 = 0$ differing by a multiple of $\partial/\partial\xi_1$. But this is $\partial/\partial x_2$ by (2.11), which gives

$$(\partial/\partial x_2)^j (\partial/\partial x_1)^i p_2 = 0 \text{ at } 0 \text{ if } i+(s+1)j < k. \tag{2.12}$$

This means that $p_2(x_1, x_2, 0)$ vanishes of weight k at 0. Now we apply Lemma 7.1 to

$$F_\varepsilon(x, \xi') = \varepsilon^{-k} p_2(\varepsilon x_1, \varepsilon^{s+1} x_2, \varepsilon^{k/2} x'', \varepsilon^{k/2}\xi')$$

where ε and x_2 are regarded as parameters and $x'' = (x_3, \cdots, x_n)$. If $b(x_1, x_2)$ is the sum of monomials of weight k in $p_2(x_1, x_2, 0)$ then Taylor expansion at $(\varepsilon x_1, \varepsilon^{s+1} x_2, 0)$ shows in view of (2.9), (2.11) that

$$F_\varepsilon(x, \xi') = b(x_1, x_2) + \varepsilon^{s-k/2} x_1{}^s \xi_2/s! + \mathcal{O}(\varepsilon^{s+1-k/2}).$$

Hence the linear form L in Lemma 7.1 can be taken to be ξ_2, so F_ε must be bounded as $\varepsilon \to 0$ if $\xi_2 = 0$. All terms in $p_2(x, \xi')$ which contain no factor ξ_2 are therefore of weight $\geqq k$ with strict inequality except for monomials in x_1 and x_2. Since the weight $k-s$ of ξ_2 exceeds $k/2$ the only other term of weight $\leqq k$ is $x_1{}^s \xi_2/s!$. Thus we conclude that

$$p_2(x, \xi') - (b(x_1, x_2) + x_1{}^s \xi_2/s!) \text{ vanishes of weight } > k;$$

where b is a polynomial of homogeneous weight k. By Lemma 7.3 we also know that this main term must satisfy (2.1). When we take $x_1 \neq 0$ and solve the equation $b(x_1, x_2) + x_1{}^s \xi_2/s! = 0$ for ξ_2, it follows that

$$\partial b/\partial x_1 - s\, b/x_1 \geqq 0 \text{ when } x_1 \neq 0. \tag{2.13}$$

In view of (2.11) there can be no term of degree s with respect to x_1 in b. If we write

$$b(x_1, x_2) = \sum_{i+(s+1)j=k} b_{ij} x_1^i x_2^j$$

the condition (2.13) can be written explicitly as

$$B(x_1, x_2) = \sum_{i+(s+1)j=k} b_{ij}(i-s)\, x_1^{i-1} x_2^j \geqq 0 . \tag{2.13$'$}$$

Note that (2.13) shows that b is divisible by x_1 if $s \neq 0$; if $s = 0$ this is also true since b contains no term with $i = s$. Also note that b is identically 0 if and only if B is. The homogeneity

$$B(-x_1, (-1)^{s+1} x_2) = (-1)^{k-1} B(x_1, x_2)$$

shows that k must be odd if b is not identically 0.

It is easy to see that B must be divisible by x_1^2 if $s \neq 0$. This is clear if $s = 1$. If $s > 1$ then the coefficient c of $x_1 x_2^j$ in b, with $j = (k-1)/(s+1)$, must be $\leqq 0$ by (2.13)$'$. Since $b(0, 1) = 0$ implies $\partial b(0,1)/\partial x_1 \geqq 0$ by Lemma 7.3 and (2.1), we must have $c \geqq 0$ so $c = 0$. It is then clear that x_1^3 must divide b since the lowest order term in B with respect to x_1 would otherwise be of odd order.

By another limiting process where ξ_2 is replaced by $A\xi_2$ and $A \to \infty$ after division by A we find that $\xi_1 + ix_1^s \xi_2$ must also satisfy (2.1) which implies that $s \neq 1$. Summing up, we have now proved

LEMMA 2.3. *Assume that* p *satisfies* (2.1) *and that* $2s < k < \infty$ *for the integers defined by* (2.2) *and* (2.3). *If* p *is replaced by its product with some non-vanishing function one can introduce canonical coordinates so that* $p_1 = \xi_1$, p_2 *is independent of* ξ_1 *and with the weights* (2.6)$'$

$$p_2(x, \xi') - b(x_1, x_2) - x_1^s \xi_2 / s!$$

vanishes of weight $k + 1/2$ *at* 0. *Here* b *is a polynomial of homogeneous weight* k *which contains no term which is of degree* s *in* x_1. *If* $s \neq 0$ *then* $s > 1$ *and* x_1^3 *divides* b. *In case* $s < \infty$ *and* $k = \infty$ *the preceding statements are true with* $b = 0$ *for any integer* $> 2s$ *instead of* k.

If $b = 0$ identically, it follows from Lemma 2.1 that all Poisson brackets of p_1 and p_2 with at most $k+1$ factors vanish at 0 in contradiction with the definition of k. In fact, there is no way then in which the factor ξ_2 can disappear. On the other hand, we shall now prove that if all the Poisson brackets of degree $k+1$ vanish then B must be identically 0. In fact, we shall prove much more by computing the left hand side of (2.4). Lemma 2.1 shows that we may replace the functions f_j, g_j by their values at 0, that is, regard them as constants. We may also replace p_2 by $x_1^s\xi_2/s! + b$. We have then

$$(2.14)\qquad (f_1H_1 + f_2H_2)(g_1p_1 + g_2p_2) = (f_1g_2 - f_2g_1)H_1p_2 = \\ = (f_1g_2 - f_2g_1)(x_1^{s-1}\xi_2/(s-1)! + \partial b/\partial x_1)\ ,$$

$$(2.15)\quad f_1H_1 + f_2H_2 = f_1\partial/\partial x_1 + f_2(x_1^s/s!\,\partial/\partial x_2 - \partial b/\partial x_2\partial/\partial\xi_2) \bmod \partial/\partial\xi_1\,.$$

The integral curve of this vector field through 0 is obtained from

$$dx_1/dt = f_1,\ dx_2/dt = f_2x_1^s/s!,\ d\xi_2/dt = -f_2\partial b/\partial x_2$$

which gives

$$x_1 = f_1t,\ x_2 = f_1^sf'_2\,t^{s+1},\ \xi_2 = -t^{k-s}f_2(k-s)^{-1}b_2(f_1, f_1^sf'_2)\,.$$

Here we have used the notation $f'_2 = f_2/(s+1)!$ and $b_j = \partial b/\partial x_j$. On this curve the right hand side of (2.14) is equal to

$$(2.16)\qquad (f_1g_2 - f_2g_1)t^{k-1}(b_1(f_1, f_1^sf'_2) - \frac{s(s+1)}{(k-s)}f_1^{s-1}f'_2\,b_2(f_1, f_1^sf'_2))\,.$$

Differentiation of the identity

$$b(tx_1, t^{s+1}x_2) = t^kb(x)$$

with respect to t gives when $t = 1$ that

$$k\, b(x) = x_1 b_1(x) + (s+1) x_2 b_2(x) .$$

Hence

$$\begin{aligned} b_1(x) - s(s+1) b_2(x) x_2/x_1(k-s) &= (k/(k-s))(b_1(x) - sb(x)/x_1) \\ &= k\, B(x)/(k-s) \end{aligned}$$

where B is defined by (2.13)′, which shows that (2.16) is equal to

$$t^{k-1}(f_1 g_2 - f_2 g_1) B(f_1, f_1{}^s f'_2) k/(k-s) .$$

Differentiating $k-1$ times with respect to t we now obtain

$$(2.17)\quad (f_1 H_1 + f_2 H_2)^k (g_1 p_1 + g_2 p_2)(0) = (f_1 g_2 - f_2 g_1) k!(k-s)^{-1} B(f_1, f_1{}^s f'_2) .$$

Note that

$$B(f_1, f_1{}^s f'_2) = f_1{}^{k-1} B(1, f'_2/f)$$

where $B(1,y)$ is a non-negative polynomial of degree at most $(k-1)/(s+1)$. A more precise discussion of the degree will follow, but first we shall examine the Hamilton fields of the Poisson brackets of p_1 and p_2 under the hypotheses in Lemma 2.3.

If $|I| < k/2$ then $H^I p_2$ vanishes at 0 of weight $\geq k - |I| > k/2$. Replacing p_2 by $b(x_1, x_2) + x_1{}^s \xi_2/s!$ changes $H^I p_2$ by a term vanishing of weight at least $k - |I| + 1/2 > (k+1)/2$. It follows that dx'' and $d\xi''$ cannot occur in $dH^I p_2$ then. Nor can dx_1 for its weight 1 is $\leq k/2$ since we have excluded the case $k = 1$. dx_2 cannot be present unless x_2 occurs in b, and then the positivity of B shows that $x_2{}^2$ must occur in b so $k \geq 2(s+1)+1$, thus $k/2 > s+1$. Hence the differentials of dp_1 and $d\,H^I p_2$ are linear combinations of $d\xi_1$ and $d\xi_2$ at 0 if $|I| < k/2$. If $|I| < s$ then $H^I p_2$ vanishes of weight $k - |I| > k - s$ at 0 so $d\xi_2$ cannot occur either. However, $H_1{}^s p_2 - \xi_2 - \partial^s b/\partial x_1{}^s$ vanishes of weight $> k-s$ so $d\,H_1{}^s p_2 = d\xi_2$ at 0. It follows that the exponent s

in Lemma 2.3 must agree with the number defined by (2.3) for any p_2 satisfying the conditions there.

We have now proved most of the following counterpart of Theorem 2.2.

THEOREM 2.4. *Let* p_1 *and* p_2 *satisfy* (2.1) *and assume that* γ *is a common zero where* $2s < k$ *if* k *and* s *are defined by* (2.2) *and* (2.3). *Then the commutators of the Hamilton fields* H_1 *and* H_2 *with less than* $1+k/2$ *factors span a two dimensional space of tangent vectors at* γ. *If* $k<\infty$ *it follows that* k *is odd and that for* $f_1, \cdots, g_2 \in C^\infty$

$$(2.19)\quad (H_{f_1p_1+f_2p_2})^k (g_1p_1+g_2p_2)(\gamma) = (f_1g_2-f_2g_1)Q(f_1,f_2)(\gamma)$$

where Q *is a homogeneous polynomial which is positive except on at most* $1+(k-3)/2(s+1)$ *lines in* $\mathbb{R}^2$. *If* (2.7) *is fulfilled then* s *is even and* Q *vanishes on at most* $(k+s+1)/2(s+1)$ *lines through* 0.

Proof. As usual we may assume that $p_1 = \xi_1$, that p_2 is independent of ξ_1 and that p_2 has the form discussed in Lemma 2.3. First we prove that if (2.7) is valid, that is, if p_2 does not change sign from $+$ to $-$ for increasing x_1, then s must be even and x_1^{s+1} divides b. Since condition (2.7) is preserved under passage to the limit it is sufficient to prove this when $p_2(x,\xi) = b(x_1,x_2) + x_1^s\xi_2/s!$. Taking ξ_2 large we see that s must be even. If for given x with $x_1 \neq 0$ we choose ξ_2 so that $b(x_1,x_2)/x_1^s + \xi_2/s! > 0$ then we know that this inequality must hold in the wide sense when x_1 is replaced by any larger number. Thus $b(x_1,x_2)/x_1^s$ must be increasing which shows that x_1^s must divide $b(x_1,x_2)$. Since no term of degree s with respect to x_1 occurs in b this proves that x_1^{s+1} divides $b(x_1,x_2)$. Note that the monotonicity of $b(x_1,x_2)/x_1^s$ as a function of x_1 completely expresses the condition (2.13).

The polynomial $B(f_1, f_1^s f_2')$ on the right hand side of (2.17) is homogeneous and non-negative of degree $k-1$ so it can only vanish on $(k-1)/2$

lines, which proves the statement on Q when $s = 0$. When $s > 1$ we write the polynomial in the form

$$f_1^{k-1} B(1, f_2'/f_1) .$$

The non-negative polynomial $B(1, y)$ is by Lemma 2.3 of degree at most $(k-3)/(s+1)$ (resp. $(k-s-1)/(s+1)$ if (2.7) is fulfilled) which means that $B(f_1, f_1^s f_2')$ can vanish on at most half that many lines in addition to the line $f_1 = 0$. This completes the proof.

There are no restrictions on b beyond those given in Lemma 2.3 and in the remarks at the beginning of the proof. First of all, if $s = 0$ then

$$p_2(x, \xi) = \xi_2 + b(x_1, x_2)$$

obviously satisfies (2.7) if the condition $\partial b/\partial x_1 \geqq 0$ in (2.13) is fulfilled. Now let $s > 1$ and

$$p_2 = x_1^s \xi_2/s! + b(x_1, x_2) ,$$

where b is divisible by x_1^3. Then (2.1) is obvious when $x_1 = 0$ and is equivalent to (2.13) when $x_1 \neq 0$. Finally, if $b(x) = x_1^s e(x)$ and s is even, $\partial e/\partial x_1 \geqq 0$, it follows that

$$p_2(x, \xi) = x_1^s(\xi_2/s! + e(x))$$

satisfies (2.7).

Let us finally make some remarks on the case where (2.7) is fulfilled but $k = \infty$ at some point γ, although this does not happen in the subelliptic case. It follows then from Theorem 2.4 that the commutators of the Hamilton fields H_1 and H_2 span a one or two dimensional space at γ. If p is analytic it follows from a version of Frobenius' theorem due to Nagano [8] that there is a one or two dimensional analytic manifold Γ through γ with tangent plane everywhere spanned by the commutators $H_{[I]} = H_{p_I}$. Since the derivatives of p_I along the vector field $H_{[J]}$ all vanish at y, the analyticity shows that $p_I = 0$ on Γ for every I. It follows that $\sigma(H_{[I]}, H_{[J]}) = \{p_I, p_J\} = 0$ in Γ so Γ is isotropic. When p is the principal symbol of an operator P it is then easy to show that Γ can carry the singularities of some u with $Pu \in C^\infty$. The Hamilton field H_p defines a natural complex structure in Γ in the two dimensional case, for it inherits the property (2.7). Thus the situation is very close to that for operators satisfying condition (P) examined in Hörmander [6].

3. *Necessary conditions for subellipticity*

It is natural to microlocalize the definition of subellipticity given in the introduction as follows (see [7, section 2] for the terminology):

DEFINITION 3.1. The properly supported pseudo-differential operator[1)] P of order m in Ω is said to be subelliptic with loss of $\delta < 1$ derivatives at $(x_0, \xi_0) \in T^*(\Omega)\setminus 0$ if

$$u \in \mathcal{D}'(\Omega),\ Pu \in H_{(s)} \text{ at } (x_0, \xi_0) \Longrightarrow u \in H_{(s+m-\delta)} \text{ at } (x_0, \xi_0)\,. \tag{3.1}$$

It is clear that this property only depends on the germ of the principal symbol p at (x_0, ξ_0) (see the introduction) and also that subellipticity at every point in $T^*(\Omega)\setminus 0$ is equivalent to subellipticity in Ω. Since the product of a subelliptic operator and an elliptic operator is again subelliptic with the same loss of derivatives, we may always assume $m = 1$ when this is convenient.

Let $K \subset \Omega$ be a compact neighborhood of x_0. Then (3.1) implies that one can find $\psi(x, D)$ homogeneous of order 0, non-characteristic at (x_0, ξ_0), such that

$$\|\psi(x, D)u\|_{(m-\delta)} \leqq C(\|Pu\|_{(0)} + \|u\|_{(m-1)}),\ u \in C_0^\infty(K)\,. \tag{3.2}$$

In fact, in the Hilbert space of all $u \in \mathcal{E}'(K) \cap H_{(m-1)}$ with $Pu \in H_{(0)}$

---

1) We assume that the symbol of P is a sum of functions homogeneous of degree $m, m-1, \cdots$.

and norm $(\|Pu\|_{(0)}^2 + \|u\|_{(m-1)}^2)^{1/2}$ the set of all u with $\|\psi(x,D)u\|_{(m-\delta)} \leqq N$ is closed, convex and symmetric. If ψ_j is a sequence such that every conic neighborhood of (x_0,ξ_0) contains the cone support of some ψ_j, then $\|\psi_j(x,D)u\|_{(m-\delta)} \leqq N$ for every u with some j and N, depending on u. By Baire's theorem it follows that we can find j and N so that this is true in a neighborhood of 0, which gives (3.2).

Conversely, if (3.2) is valid and $u \in H_{(m-1)}$ has support in the interior of K, $Pu \in H_{(0)}$, then an application of (3.2) to the regularizations of u shows that $\psi(x,D)u \in H_{(m-\delta)}$. It follows that $u \in H_{(m-\delta)}$ at (x_0,ξ_0) whenever $u \in H_{(m-1)}$ and $Pu \in H_{(0)}$ at (x_0,ξ_0), for we can always replace u by $\psi_1(x,D)u$ with ψ_1 of order 0 and cone supp ψ_1 close to (x_0,ξ_0) to reduce to the case considered first. Now if $Pu \in H_{(s)}$ and $u \in H_{(s+m-1)}$ at (x_0,ξ_0) it follows that $u \in H_{(s+m-\delta)}$ at (x_0,ξ_0), for if E_s has symbol $|\xi|^s$ and $v = E_s u$, then $E_s P E_{-s} v \in H_{(0)}$ and $v \in H_{(m-1)}$ at (x_0,ξ_0). The operator $E_s P E_{-s}$ has the same principal symbol as P, and (3.2) remains valid with some other constants if the lower order terms of P are changed. Hence $v \in H_{(m-\delta)}$ and $u \in H_{(s+m-\delta)}$ at (x_0,ξ_0), so the usual inductive argument shows that (3.1) is fulfilled. Thus our problem is to analyze the conditions for validity of the estimate (3.2).

Choose $\phi \in C_0^\infty(R^{2n})$ with support near the origin, and apply (3.2) with u replaced by

$$\phi(x-x_0, \lambda^2 D-\xi_0)u\ ,$$

where we assume now that ξ_0 is a unit vector and λ is small. The symbol used here is in a bounded subset of $S^0_{1,0}$. We assume that $m = 1$ and that the symbol p of P is homogeneous of degree 1. In (3.2) we need only take the first term in the symbol calculus into account then, for the others may be included in the error term. Now we make a translation of the x,ξ variables and a symplectic dilation, which introduces the symbol

$$\phi(\lambda x, \lambda \xi) p(x_0 + \lambda x, \lambda^{-2}\xi_0 + \xi/\lambda) = \lambda^{-2} q(\lambda x, \lambda \xi)$$

where

$$q(x, \xi) = \phi(x, \xi) p(x_0 + x, \xi_0 + \xi) .$$

(We take $\phi = 1$ near 0 so this is equal to $p(x_0 + x, \xi_0 + \xi)$ near 0.) With $\phi_1(x, \xi) = \phi(x, \xi) |\xi_0 + \xi|^{1-\delta} \psi(x_0 + x, \xi_0 + \xi)$ we obtain

$$(3.3) \qquad \lambda^{2(\delta-1)} \|\phi_1(\lambda x, \lambda D) u\| \leqq C(\lambda^{-2} \|q(\lambda x, \lambda D) u\| + \|u\|), \ u \in \mathcal{S} ,$$

where the norms are L^2 norms.

Conversely, it is easy to return from (3.3) to (3.2) for a suitable $\psi(x, D)$. We just have to replace u by $\phi_2(\lambda x, \lambda D) u$ where ϕ_2 is another function in C_0^∞, return to the original variables, square and integrate with respect to $d\lambda/\lambda$ from 0 to 1. We omit the details.

We are thus led to the problem of determining all $q \in C_0^\infty(R^{2n})$ such that (3.3) is valid for some $\phi_1 \in C_0^\infty$ which is 1 in a neighborhood of 0. It is clear that the answer can only depend on the germ of q at 0. Note that this is some generalization of our original problem since we now drop the conditions related to the homogeneity of p. This makes it possible to apply non-homogeneous canonical transformations:

LEMMA 3.2. *If* (3.3) *is valid and* χ *is a* C^∞ *canonical transformation keeping* 0 *fixed, then* (3.3) *remains valid (with some other* ϕ_1 *and* C) *if* q *is replaced by* $q \circ \chi$.

Proof. This is essentially the Egorov theorem occurring in [7, Lemma 6.1] although it is not tied to a homogeneous canonical transformation here. It suffices to prove the lemma when χ has a generating function S, for every canonical transformation is the composition of two such transformations. Note that $S = S' = 0$ at 0 and that

$$S_\lambda(x, \eta) = S(\lambda x, \lambda \eta)/\lambda^2$$

corresponds to the canonical transformation $\chi_\lambda(x,\xi) = \lambda^{-1}\chi(\lambda x, \lambda\xi)$. (A function of the form S_λ occurs of course if one starts from a phase function which is homogeneous of degree 1 and changes variables as above.) Let $\phi_2 \epsilon C_0^\infty$ have support near 0 and be 1 in some neighborhood of 0, and introduce the Fourier integral operator

$$T_\lambda u(x) = \int \phi_2(\lambda x, \lambda\eta)(\exp iS_\lambda(x,\eta))\hat{u}(\eta)\,d\eta \ .$$

As usual one obtains with a_λ bounded in $\mathcal{S}$ as $\lambda \to 0$

$$T_\lambda^* T_\lambda u(x) = \int a_\lambda(\lambda x, \lambda\eta) e^{i\langle x,\eta\rangle}\hat{u}(\eta)\,d\eta \ ,$$

and $a_\lambda(0)$ has a limit which is not 0. In particular there is a bound for the norm of T_λ, and $T_\lambda^* T_\lambda$ converges to a constant $\neq 0$ when $\lambda \to 0$. Moreover, if $q' = q \circ \chi$ then

$$q(\lambda x, \lambda D)\,T_\lambda = T_\lambda q'(\lambda x, \lambda D) + \lambda^2 R_\lambda$$

where R_λ is of the same form as T_λ, thus uniformly bounded in L^2. Replacing u by $T_\lambda u$ in (3.3) one therefore obtains an estimate of the same form with q replaced by q', which concludes our sketch of the proof.

It is also clear that (3.3) remains valid if q is multiplied by a non-vanishing C^∞ factor. The results proved in section 2 will therefore lead to necessary conditions for subellipticity once the following lemma is proved:

LEMMA 3.3. *(3.3) cannot hold with $\phi_1(0) \neq 0$ and $\delta < 1$ if near 0*

$$q(x,\xi) = \xi_1 - ix_1^{\,k},\ k > 0 \ \textit{odd}\ .$$

If q vanishes of weight k at 0 when x_j, ξ_j are given positive weights m_j, μ_j with $m_j + \mu_j = k+1$, then (3.3) implies that

(3.5) $$\delta \geqq k/(k+1)\ .$$

Proof. Let us first assume that $q(x,\xi) = (\xi_1 - ix_1^k)\phi_2(x,\xi)$ for some $\phi_2 \in C_0^\infty$. Then

$$q(\lambda x, \lambda D)u = \phi_2(\lambda x, \lambda D)((\lambda D_1 - i(\lambda x_1)^k)u) + \\ + \sum_1^k i^{1-j}\lambda^{2j}\binom{k}{j}(\lambda x_1)^{k-j}\phi_2^{(j)}(\lambda x, \lambda D)u$$

where $\phi_2^{(j)}(x,\xi) = \partial^j\phi_2(x,\xi)/\partial\xi_1{}^j$. All the terms in the sum can be bounded by $\lambda^2\|u\|$, so (3.3) implies

(3.3)′ $$\lambda^{2(\delta-1)}\|\phi_1(\lambda x, \lambda D)u\| \leqq C(\|(D_1 - i\lambda^{k-1}x_1^k)u\|/\lambda + \|u\|)\ .$$

We may assume that $\phi_1(0) = 1$ and write

$$\phi_1(x,\xi) = 1 + \sum x_j f_j(x,\xi) + \sum \xi_j g_j(x,\xi)$$

where f_j and g_j have bounded derivatives of all orders. Then

$$\phi_1(\lambda x, \lambda D)u = u + \sum f_j(\lambda x, \lambda D)(\lambda x_j u) + \sum g_j(\lambda x, \lambda D)(\lambda D_j u) \\ + i\lambda^2 \sum f_{j'}(\lambda x, \lambda D)u$$

where $f_{j'} = \partial f_j/\partial\xi_j$. Thus it follows from (3.3)′ that

(3.3)″ $$\lambda^{2(\delta-1)}\|u\| \leqq C(\|(D_1 - i\lambda^{k-1}x_1^k)u\|/\lambda + \|u\| + \\ + \lambda^{2(\delta-1)}\sum(\|\lambda x_j u\| + \|\lambda D_j u\|))\ .$$

Now we take

$$u(x) = v(x_2, \cdots, x_n)\exp -(\lambda^{k-1}x_1^{k+1}/(k+1))$$

with $v \in C_0^\infty(R^{n-1})$ not identically 0. When we introduce

$$t = x_1\lambda^{(k-1)/(k+1)}$$

and note that

$$\lambda x_1 = \lambda^{2/(k+1)}t,\ \lambda\partial/\partial x_1 = \lambda^{2k/(k+1)}\partial/\partial t$$

it becomes clear that the right hand side of (3.3)″ is smaller than the left hand side when λ is small. This proves the first part of the lemma.

To prove the second part of the lemma we first make a symplectic dilation, replacing x_j by $\tau^{-\sigma_j}x_j$ and D_j by $\tau^{\sigma_j}D_j$ where $\tau^{k+1} = \lambda^2$,

$$\sigma_j = \mu_j - (k+1)/2 = (k+1)/2 - m_j .$$

This makes $\lambda\tau^{-\sigma_j} = \tau^{m_j}$ and $\lambda\tau^{\sigma_j} = \tau^{\mu_j}$, so we obtain with obvious notations that (3.3) is equivalent to

$$(3.4)\quad \tau^{(k+1)(\delta-1)}\|\phi_1(\tau^m x, \tau^\mu D)u\| \leqq C(\tau^{-(k+1)}\|q(\tau^m x, \tau^\mu D)u\| + \|u\|),\ u \in \mathcal{S}.$$

We can write by Taylor's formula

$$q(x,\xi) = \sum q_{\alpha\beta}(x,\xi)x^\beta\xi^\alpha$$

where $q_{\alpha\beta} \in C_0^\infty$, the sum is finite and $\langle\alpha,\mu\rangle + \langle\beta,m\rangle \geqq k$ for every term in the sum. Since

$$q(\tau^m x, \tau^\mu D) = \sum \tau^{\langle\alpha,\mu\rangle+\langle\beta,m\rangle}x^\beta q_{\alpha\beta}(\tau^m x, \tau^\mu D)D^\alpha ,$$

the right hand side of (3.4) is bounded by C/τ for any fixed u. On the left hand side

$$\phi_1(\tau^m x, \tau^\mu D)u \to \phi_1(0)u,\ \tau \to 0 ,$$

so we conclude that

$$(k+1)(\delta-1) \geqq -1 ,$$

that is, $\delta \geqq k/(k+1)$ as claimed.

We have now proved everything that is required for the proof of the necessity in the following main theorem of Egorov:

THEOREM 3.4. *The pseudo-differential operator* P *with principal symbol* p *is subelliptic at* $(x_0, \xi_0) \in T^*(\Omega)\setminus 0$ *with loss of* $\delta < 1$ *derivatives if and only if there is a neighborhood* V *of* (x_0, ξ_0) *such that*
(i) *for every* $\gamma \in V$

$$(H_{\mathrm{Re}\, zp})^j \,\mathrm{Im}\, zp\, (\gamma) \tag{3.5}$$

is different from 0 *for some* $z \in \mathbb{C}$ *and some* $j \leqq \delta/(1-\delta)$.
(ii) (3.5) *is non-negative if* j *is odd and is the smallest integer such that* (3.5) *is not identically* 0.

Proof of the necessity. Subellipticity implies an estimate (3.3) (which conversely implies subellipticity at every point in a neighborhood of (x_0, ξ_0)). If $p = 0$ but $\{\mathrm{Re}\, p, \mathrm{Im}\, p\} < 0$ at a point near (x_0, ξ_0) then the product of p and some non-vanishing function is of the form $x_1 - i\xi_1$ with a coordinate system centered there. This follows from the simpler analogue of [7, Theorem 7.1] where homogeneity conditions are dropped. Thus Lemma 3.2 and Lemma 3.3 with $k = 1$ show that this cannot happen, so (2.1) must be valid near (x_0, ξ_0). The full condition (ii) follows by the discussion following (2.7) if one uses Lemma 3.3 for arbitrary odd k (see also [7, Theorem 7.3]). From Lemma 3.2 and the second part of Lemma 3.3 we obtain $\delta \geqq k/(k+1)$ for the integer k defined by (2.2) if we recall Lemma 2.3 and the proof of Theorem 2.2. Thus $k \leqq \delta/(1-\delta)$ as claimed.

REMARK. Since all Poisson brackets of Re p and Im p are homogeneous functions, the condition (i) implies that H_p is not proportional to the radial vector $\langle \xi, \partial/\partial\xi \rangle$ at any zero of p. After multiplication by an elliptic factor this allows one to transform p locally by a *homogeneous* canonical transformation to the form

(3.6) $$p(x,\xi) = \xi_1 + ip_2(x,\xi') .$$

Since subellipticity is invariant under conjugation by elliptic Fourier integral operators one may always assume that p has the form (3.6) in proving the sufficiency of the conditions in Theorem 3.4.

The purpose of the remaining part of the paper is to prove the sufficiency in Theorem 3.4. In fact, we shall prove somewhat more by showing that (3.3) is valid if q is obtained as there from a function p satisfying the hypotheses of Theorem 3.4. There is nothing to prove when $q(0) \neq 0$, and when $q(0) = 0$ we may assume in view of Lemma 3.2 that

$$q(x,\xi) = \xi_1 + iq_2(x,\xi') \text{ for } |x|^2 + |\xi|^2 < c^2 .$$

Here $\xi' = (\xi_2, \cdots, \xi_n)$. Then

$$\lambda^{-2} q(\lambda x, \lambda\xi) = \lambda^{-1}\xi_1 + i\lambda^{-2} q_2(\lambda x_1, \lambda x', \lambda\xi')$$

which makes it natural to introduce λx_1 and ξ_1/λ as new variables. This symplectic change of variables reduces the symbol to

$$\xi_1 + i\lambda^{-2} q_2(x_1, \lambda x', \lambda\xi'), \ |x_1|^2 + |\lambda^2\xi_1|^2 + |\lambda x'|^2 + |\lambda\xi'|^2 < c^2 .$$

Now we change the notation and write

$$q(x,\xi') = \lambda^{-2} q_2(x_1, \lambda x', \lambda\xi') .$$

Thus the parameter λ is suppressed and just remains in the estimates

$$|D_\xi^\alpha D_x^\beta q(x,\xi')| < C_{\alpha\beta}\lambda^{|\alpha'+\beta'|-2} ;$$

we assume that they are valid when $|x_1| < 1$ and $|(x',\xi')| < 1/\lambda$ which is no real restriction. In view of (2.7) q does not change sign from $+$ to $-$ for increasing x_1 then, and moreover condition (i) above means that some Poisson bracket of at most $k+1$ factors ξ_1 and q has absolute value $\geqq c_0\lambda^{-2}$ at any such point. These properties will be our starting point in section 4. After a local study of q in section 4 and a study of localizations of $D_1 + i\lambda^{-2} q(x_1, \lambda x', \lambda D')$ in section 5, an estimate which implies (3.3) will be proved for this operator in section 6.

4. *Local properties of the principal symbol*

Let $q(x, \xi')$ be a C^∞ function satisfying

$$(4.1) \qquad |D_\xi^\alpha D_x^\beta q(x,\xi')| \leqq C_{\alpha\beta}\lambda^{|\alpha'+\beta'|-2};\ |x_1| < 1,\ |(x',\xi')| < 1/\lambda .$$

Here $C_{\alpha\beta}$ are fixed constants, λ is small but q and λ are allowed to vary. By ξ' for example we mean the vector $(\xi_2, \cdots, \xi_n)$. Sometimes we shall write $p_1 = \xi_1$ and $p_2 = q$ to have the symmetric notations of section 2 available. Our second condition is

(4.2) $q(x,\xi')$ does not change sign from $+$ to $-$ for increasing x_1 .

The third and last condition is that for some fixed C and k

$$(4.3) \qquad \lambda^{-2} \leqq C \sum_{|I| \leqq k+1} |p_I(x,\xi)| .$$

Here it is of course advantageous to put $\xi_1 = 0$, that is, drop the term ξ_1 on the right hand side.

Following Egorov we shall now introduce a quite precise measure for the size of the commutators, which is analogous to the definition (2.2) of the integer k,

$$(4.4) \qquad M(x,\xi') = \max_{|I| \leqq k+1} |p_I(x,\xi)/\rho|^{1/|I|},\ \xi_1 = 0 .$$

Here ρ is a large parameter. (Actually Egorov uses about k^2 parameters.) The purpose of ρ is to guarantee a large constant in the left hand side of an estimate which allows one to cancel some error terms on the right hand side. (See the proof of Lemma 5.3 for a quite elementary example of how this works.) We shall ultimately fix ρ large but independent of λ so that the desired estimates follow for small enough λ.

By (4.1) and (4.3) we have

$$C_1(\lambda^{-2}/\rho)^{1/(k+1)} \leqq M(x,\xi') \leqq C_2\lambda^{-2}/\rho \,. \tag{4.5}$$

Here and in what follows the constants are independent of λ and ρ. Note that (4.5) shows that the size of M would not change very much if k were increased in (4.4), so the definition of M is essentially independent of k if k is only large enough to make (4.3) valid.

We shall analyze the behavior of q near an arbitrary point but to simplify notations it is convenient to take it to be the origin. The definition of $M = M(0)$ means in particular that

$$|D_{x_1}{}^j q(0)| \leqq \rho M^{j+1}, \; j \leqq k \,, \tag{4.6}$$

and where (4.1) is valid we have by (4.5)

$$|D_{x_1}{}^j q| \leqq C_j \rho M^{k+1} \tag{4.7}$$

which is a much better estimate when $j > k$. If we set

$$F(t, y', \eta') = q(t/M, y'\sqrt{\rho M}, \eta'\sqrt{\rho M})/\rho M$$

and note that $\rho M \leqq C_2\lambda^{-2}$ it follows from (4.6), (4.7) and Taylor's formula that F satisfies (7.1) after division by a constant. A constant scale change in the (x',ξ') variables makes F defined in Ω. (7.3) follows from (4.1) since the right hand side in (4.1) is constant when $|\alpha' + \beta'| = 2$. Since (4.1), (4.5) give

$$|d_{x'\xi'} D_{x_1}{}^j q| \leqq C_j(M^{k+1}\rho)^{1/2} \,, \tag{4.8}$$

we obtain

$$|d_{y'\eta'} D_t{}^j F| \leqq C_j \quad \text{if } j \geqq k/2$$

which proves (7.2) with k replaced by any such value of j. Moreover, there is a uniform bound for $D_{\eta'}^{\alpha'} D_t^{\beta_1} D_{y'}^{\beta'} F$ if $|\alpha' + \beta'| \geqq 2$. After a symplectic orthogonal transformation of the $y'\eta'$ variables it follows therefore from Lemma 7.1 that

$$F(t, y', \eta') - \eta_2 \partial F(t, 0)/\partial \eta_2$$

and any derivative has a uniform bound. Returning to the original variables, we obtain, with $\tilde{q}$ differing from q by a symplectic orthogonal transformation and with new $C_{\alpha\beta}$, provided that $|x_1 M| < 1, |(x', \xi')| < C\sqrt{\rho M}$,

$$(4.9) \quad |D_\xi^\alpha D_x^\beta (\tilde{q}(x, \xi') - \xi_2 \partial \tilde{q}(x_1, 0)/\partial \xi_2)| \leqq C_{\alpha\beta} \rho M^{\beta_1+1} (\rho M)^{-|\alpha' + \beta'|/2} .$$

As in section 2 we shall have to distinguish two cases depending on the size of the linear term in (4.9). The following simple analogue of Lemma 2.1 will indicate how the division should be made.

LEMMA 4.1. *If* $p = p_1 + ip_2$ *and for some positive numbers* $A_1, \cdots, A_n, B_1, \cdots, B_n$ *we have*

$$(4.10) \qquad |D_\xi^\alpha D_x^\beta p(0)| < C_{\alpha\beta} \rho K A^\alpha B^\beta ,$$

and if $A_j B_j \rho \leqq 1$ *for every* j, *then*

$$(4.11) \qquad |p_I(0)| \leqq C_I \rho K^{|I|}$$

where each constant only depends on a finite number of the constants $C_{\alpha\beta}$ *in* (4.10).

Proof. By induction it follows immediately that p_I has estimates of the form (4.10) with K replaced by $K^{|I|}$.

REMARK 1. Lemma 4.1 generalizes Lemma 2.1 for if p vanishes of weight k as in Lemma 2.1 then

$$|p^{(\alpha)}_{(\beta)}(\varepsilon^m x, \varepsilon^\mu \xi)| < C\,\varepsilon^{k-<\alpha,\mu>-<\beta,m>} .$$

With $A_j = \varepsilon^{-\mu_j}$, $B_j = \varepsilon^{-m_j}$, $\rho = \varepsilon^{k+1}$ and $K = 1/\varepsilon$ in Lemma 4.1 we obtain

$$p_I(\varepsilon^m x, \varepsilon^\mu \xi) = \mathcal{O}(\varepsilon^{k+1-|I|})$$

which means that p_I vanishes of weight $k+1-|I|$.

REMARK 2. Note that if $A_j B_j \rho << 1$ for a certain j, then the terms in p_I involving some derivative with respect to x_j, ξ_j will have a much smaller bound than (4.11).

Let ε be a small positive number such that

$$1 \leqq M\rho\,\varepsilon^2 . \tag{4.12}$$

In view of (4.5) this is true with a wide margin for small λ if we take $\varepsilon = \lambda^\kappa$ with $0 < \kappa < 1/(k+1)$, and this choice will be made in the following division in cases.

CASE I. Assume first that

$$|d_{x'\xi'} D_{x_1}{}^j q(0)| \leqq \varepsilon\rho M^{j+1},\ j \leqq k/2 . \tag{4.13}$$

The right hand side is at least $M^j\sqrt{M\rho}$ so when $j > k/2$ a better bound is always obtained from (4.8). Hence it follows from (4.9), (4.13), (4.8) and Taylor's formula that

$$|D^\alpha_\xi D^\beta_x q(x,\xi')| \leqq C_{\alpha\beta}\,\rho M^{\beta_1+1} \varepsilon^{|\alpha'+\beta'|},\ \text{if}\ |x_1 M| < 1,\ |(x',\xi')| < 1/\varepsilon , \tag{4.14}$$

for such an estimate is invariant under orthogonal transformations. This is of course also an immediate consequence of (4.6), (4.7), (4.13), (4.8), and (4.1) when $|\alpha'+\beta'| = 0, 1$ or $|\alpha'+\beta'| > 1$ respectively. In view of Lemma 4.1 and Remark 2 following it we now conclude, taking $A_1 = 1/M\rho$, $B_1 = M$, $A_j = B_j = \varepsilon$ otherwise, $K = M$, that (4.4) can at most change by a fixed factor if the Poisson brackets are taken just with respect to x_1, ξ_1, for $\varepsilon^2\rho << 1$. Thus it follows for small λ that

$$1 \leqq C \sum_{j \leqq k} |D_{x_1}{}^j q(0)|/\rho M^{j+1} . \tag{4.15}$$

In view of (4.6), (4.7) this means that $q(t/M, 0, 0)/\rho M$ will be very close to a polynomial of degree k with uniformly bounded coefficients and some coefficient bounded away from 0. In addition (4.14) gives

$$|q(t/M, x', \xi') - q(t/M, 0, 0)|/\rho M < C\varepsilon R_1 \quad \text{if} \quad |(x', \xi')| < R_1, |t| < 1 ,$$

and εR_1 is small. We choose $R_1 = \lambda^{-\kappa_1}$ with $0 < \kappa_1 < \kappa$ to guarantee this, and we define our standard neighborhood of 0 by

$$\omega_0 = \{(x_1, x', \xi'); |x_1| < 1/M, |x'|^2 + |\xi'|^2 < R_1^2\} .$$

Let $\phi \in C_0^\infty(-1, 1)$ be equal to 1 in $(-3/4, 3/4)$, $0 \leqq \phi \leqq 1$ everywhere, and set

$$\phi_0(x, \xi') = \phi(Mx_1)\phi((|x'|^2 + |\xi'|^2)/R_1^2) .$$

In section 6 we shall use the corresponding pseudo-differential operator in localizing our estimates. It will then be important that

$$C^{-1}M(0) \leqq M(x, \xi') \leqq C M(0) \quad \text{if} \quad (x, \xi') \in \omega_0 .$$

The estimate from above follows from (4.14) and Lemma 4.1. To prove the lower bound it suffices to show that (4.15) remains valid in ω_0. When $(x', \xi') = 0$ this follows from Taylor's formula and (4.7), with $j = k+1$, and for general (x', ξ') the same conclusion follows for small enough λ from (4.14). Thus we have also

$$M(x, \xi') \leqq C \max_{j \leqq k} |D_{x_1}{}^j q(x, \xi')/\rho|^{1/(j+1)} \quad \text{if} \quad (x, \xi') \in \omega_0 .$$

CASE II. Assume now that (4.13) is not fulfilled. Choose $s \leqq k/2$ so that with $q^{(j)} = \partial^j q/\partial x_1{}^j$

$$|d_{x'\xi'} q^{(s)}(0)| = a \tag{4.16}$$

$$|d_{x'\xi'} q^{(j)}(0)| \leqq a\, M^{j-s} \quad \text{for} \quad j \leqq k/2\,. \tag{4.17}$$

Then (4.1) and the fact that (4.13) is not valid gives

$$\varepsilon \rho\, M^{s+1} < a \leqq C/\lambda\,. \tag{4.18}$$

In view of (4.8) we have then for every j

$$|D_{x_1}{}^j \partial \tilde{q}(x_1, 0)/\partial \xi_2| \leqq C_j\, a\, M^{j-s} = C_j\, \rho\, M^{j+1} A_2,\ |x_1 M| < 1\,, \tag{4.17$'$}$$

where

$$A_2 = a/\rho\, M^{s+1}\,. \tag{4.19}$$

The estimate is sharp when $j = s$. From (4.9) we can therefore obtain an estimate of the form (4.10) with $B_2 = A_3 = B_3 = \cdots = 1/\sqrt{\rho M}$ and $K = M$. However, $\rho\, A_2 B_2 = a/M^{s+1}\sqrt{\rho M}$ so Lemma 4.1 is not applicable if $a > M^{s+1}\sqrt{\rho M}$. In that case we shall therefore just as in section 2 replace the orthogonal symplectic transformation which led from q to $\tilde{q}$ by a non-linear canonical transformation.

Thus assume for the moment that (4.16), (4.17) and

$$M^{s+1}\sqrt{\rho M} < a \leqq C/\lambda \tag{4.18$'$}$$

are fulfilled. We shall then make a canonical transformation such that $q^{(s)}(0, x', \xi')$ becomes a linear function of ξ_2. Since $\lambda a < C$ the function

$$Q(x', \xi') = a^{-2}(q^{(s)}(0, ax', a\xi') - q^{(s)}(0, 0, 0))$$

is in a bounded subset of C^∞. In fact, $Q(0,0) = 0$, the length of the gradient is 1 at 0 by the definition of a, and the derivatives of order $m \geqq 2$ can be estimated by $(a\lambda)^{m-2}$. It follows that it is possible to choose a canonical transformation χ belonging to a bounded set in C^∞ such that

$$Q \circ \chi(x', \xi') = \xi_2$$

in a fixed neighborhood of 0. Changing back to the original scales gives the canonical transformation

$$\chi_a(x', \xi') = a\chi(x'/a, \xi'/a) ,$$

and changing our earlier definition of $\tilde{q}$ to

$$\tilde{q}(x, \xi') = q(x_1, \chi_a(x', \xi'))$$

we obtain

$$\tilde{q}^{(s)}(x, \xi') = a\xi_2 + \tilde{q}^{(s)}(0,0) \quad \text{when} \quad x_1 = 0, |(x', \xi')| < ca . \tag{4.20}$$

We shall now show that this leads to a better estimate than (4.9). The reason is that (4.20) and the invariance of Poisson brackets under canonical transformations gives as in the proof of (2,12) that

$$(a\partial/\partial x_2)^j (\partial/\partial x_1)^i \tilde{q}(0) = ((\mathrm{ad}\,\partial/\partial x_1)^s H_q)^j (\partial/\partial x_1)^i q(0) .$$

By the definition of M the right hand side can be estimated by $\rho M^{1+i+j(s+1)}$ so we obtain

$$|a^j (\partial/\partial x_2)^j (\partial/\partial x_1)^i \tilde{q}(0)| \leqq C_{ij} M^{j(s+1)} \rho M^{i+1} . \tag{4.21}$$

If we introduce

$$B_2 = M^{s+1}/a , \tag{4.22}$$

noting incidentally that $A_2 B_2 \rho = 1$ as required in Lemma 4.1, this means that we have bounds for the derivatives of $\tilde{q}(x_1/M, x_2/B_2, 0)/M\rho$ at 0. Before examining the bounds for the other derivatives we pause to discuss the preceding argument.

In the derivation of (4.21) we used only a small part of (4.20) namely that the x_2 axis $x_2 = at$, $x_3 = \cdots = \xi_n = 0$, is an integral curve of the

Hamilton field of $\tilde{q}^{(s)}(0, x', \xi')$. It is possible to place the integral curve Γ of the Hamilton field of $q^{(s)}(0, x', \xi')$ in this position by a very simple canonical transformation. First note that Γ is of the form $t \to a\psi(t/a)$ where ψ belongs to a bounded subset of $C^\infty(-c, c)$ and $|\psi'(0)|$ has a fixed lower bound. (In terms of the preceding notations we have $\psi(t) = \chi(t, 0, \cdots, 0)$.) After a suitable choice of labels for the coordinates and possibly an interchange of x' and ξ' by the canonical transformation $(x', \xi') \to (\xi', -x')$, we can assume that the x_2 coordinate of $\psi'(0)$ is one with maximal absolute value. We can then take x_2 as parameter so that Γ is given by

$$x_2 = t,\ x_3 = f_3(t), \cdots, \xi_2 = g_2(t), \cdots$$

where $f_3(at)/a, \cdots$ belong to a bounded set in $C^\infty(-c', c')$ for some fixed constant c'. Now another canonical transformation $\tilde{\chi}_a$ which also pulls Γ back to the x_2 axis is given by

$$\tilde{\chi}_a(x', \xi') = (x_2, x_3 + f_3(x_2), \cdots, \xi_2 + g_2(x_2) + G(x', \xi'), \cdots, \xi_n + g_n(x_2))$$

where the additional term G given by

$$G(x', \xi') = \sum_3^n (g_j'(x_2)x_j - f_j'(x_2)\xi_j)$$

is included to make the transformation canonical. Note the linearity in all variables except x_2. It is clear that we have as good bounds for its derivatives as for those of our original χ_a, and our new choice will make it easier to find an operator corresponding to the canonical transformation.

We now return to the estimation of the other derivatives of $\tilde{q}$, assuming for the time being only the uniform bounds for $\chi_a(ax', a\xi')/a$. First we derive some crude bounds for the derivatives. If we compute

$$D_{\xi'}^{\alpha'} D_{x'}^{\beta'} \tilde{q}^{(j)}(x, \xi') = D_{\xi'}^{\alpha'} D_{x'}^{\beta'} q^{(j)}(x_1, \chi_a(x', \xi'))$$

the terms where $q^{(j)}$ is differentiated $i \geqq 2$ times can be estimated by

$$C\lambda^{i-2}\, a^{i-|\alpha'+\beta'|} \leqq C'\, a^{2-|\alpha'+\beta'|}$$

if we use (4.1) and note that there will be a product of i derivatives of components of χ_a with total order of differentiation $|\alpha'+\beta'|$; we have of course $|D^\alpha \chi_a| \leqq C_\alpha a^{1-|\alpha|}$. Furthermore, the terms where $q^{(j)}$ is only differentiated once can be estimated by

$$C(|d_{x'\xi'} q^{(j)}(x_1,0)| + a)\, a^{1-|\alpha'+\beta'|}$$

since the second order derivatives of $q^{(j)}$ with respect to (x',ξ') are bounded. As already observed in the proof of (4.17)′ we have by (4.17), (4.8)

$$|d_{x'\xi'} q^{(j)}(x_1,0)| \leqq a\, C_j\, M^{j-s},\ |x_1 M| < 1\,.$$

We omit the factor M^{-s} and obtain when $|x_1 M| < 1$, $|(x',\xi')| < ca$

$$|D_\xi^\alpha D_x^\beta \tilde{q}(x,\xi')| \leqq C_{\alpha\beta} M^{\beta_1}\, a^{2-|\alpha'+\beta'|} \quad \text{if } |\alpha'+\beta'| \neq 0\,. \tag{4.23}$$

A better estimate is sometimes obtained if one just observes that a^2 and $a\, d_{x'\xi'} q^{(j)}$ can be bounded by $C/\lambda^2 \leqq C_1 M^{k+1}\rho$. This gives

$$|D_\xi^\alpha D_x^\beta \tilde{q}(x,\xi')| \leqq C'_{\alpha\beta} M^{k+1}\rho\, a^{-|\alpha'+\beta'|} \tag{4.23$'$}$$

which is also true if $|\alpha'+\beta'| = 0$. (See also (4.7).) In particular (4.23)′ is a much better estimate than (4.21) if $j(s+1)+i > k$, and it is not only valid at 0. Hence (4.21) leads to uniform bounds for $\tilde{q}(x_1/M, x_2/B_2, 0)/\rho M$ and all of its derivatives when $|x_1| < 1$ and $|x_2| < 1$.

Since $a >> (\rho M)^{1/2}$ we can apply Lemma 7.1 to

$$F(t,x_2,y) = (M\rho)^{-1}\, \tilde{q}(t/M, x_2/B_2, x''\sqrt{\rho M}, \xi'\sqrt{\rho M})$$

where $x'' = (x_3,\cdots,x_n)$ and $y = (x'',\xi')$. We regard x_2 as a parameter at first. By the preceding observations and (4.7) the condition (7.1) is fulfilled, (7.2) follows again from (4.8), and (7.3) is now a consequence of

(4.23). Since $d\tilde{q}^{(s)}(0,x_2,0) = a\, d\xi_2 \mod dx_1$, we have

$$d_y\, \partial^s F(0,x_2,0)/\partial t^s = a\, d\xi_2/M^s\sqrt{\rho M}$$

while (4.17) and (4.8) give

$$|d_y F(t,0,0)| < Ca/M^s\sqrt{\rho M},\ |t| < 1\ . \tag{4.24}$$

By (4.23) we have

$$|d_y F(t,0,0) - d_y F(t,x_2,0)| < 1/B_2\sqrt{\rho M} = a/M^{s+1}\sqrt{\rho M}$$

so for small λ it follows that

$$|d_y F(t,x_2,0)| < 2Ca/M^s\sqrt{\rho M},\ |t| < 1,\ |x_2| < 1\ . \tag{4.24$'$}$$

Hence Lemma 7.1 and the remark after it give a uniform bound for

$$F(t,x_2,y) - \xi_2\, \partial F(t,x_2,0)/\partial\xi_2\ .$$

We have a uniform bound for all derivatives of high order if we use (4.23), (4.23)′ and observe that $\sqrt{\rho M}/a$ and $1/B_2 a$ are smaller than a positive power of λ. Thus we obtain bounds for all derivatives, and returning to the original variables we conclude that

$$|D_\xi^\alpha D_x^\beta(\tilde{q}(x,\xi') - \xi_2\, \partial\tilde{q}(x_1,x_2,0)/\partial\xi_2)| \leq C M^{\beta_1+1} \rho B_2^{\beta_2} (M\rho)^{-|\alpha'+\beta''|/2} \tag{4.25}$$

when $|x_1|M < 1$ and $|(x'',\xi')| < (\rho M)^{1/2}, |x_2| < 1/B_2$.

For the linear term in ξ_2 we have as in (4.17)′

$$B_2 |D_{x_1}{}^j\, \partial\tilde{q}(x_1/M,0)/\partial\xi_2| \leq C_j M,\ |x_1| < 1\ , \tag{4.26}$$

for $M/B_2 = a/M^s$. In addition (4.16) shows that some of the bounds (4.26) is precise, so $B_2/M\, \partial\tilde{q}(x_1/M,0)/\partial\xi_2$ is essentially a normalized polynomial in x_1 of degree $\leq k/2$. By Taylor's formula and (4.23)

$$(4.27)\quad B_2|D_x^\beta \partial(\tilde{q}(x_1/M,0)-\tilde{q}(x_1/M,x_2,0))/\partial\xi_2| \leqq C_\beta\, a^{-\beta_2},\ |x_1|<1,\ |x_2B_2|<1,$$

so $\partial\tilde{q}(x_1/M, x_2, 0)/\partial\xi_2$ is almost independent of x_2.

From (4.25), (4.26) and (4.27) we obtain with $B_1 = M$, $A_j = B_j = 1/\sqrt{\rho M}$ when $j > 2$

$$(4.28)\quad |D_\xi^\alpha D_x^\beta \tilde{q}(x,\xi')| \leqq C_{\alpha\beta N}\,\rho M A^\alpha B^\beta, \text{ if } |x_1 M|<1, |x_2B_2|<1,\ |\xi_2A_2|<N,\ |(x'',\xi'')| \leqq \sqrt{\rho M},$$

where N is a fixed but arbitrary constant. In fact, the right hand side of (4.25) has a bound of this form since $A_2 > \varepsilon > 1/\sqrt{M\rho}$ by (4.18). By (4.26), (4.27) and since $B_2 = M^{s+1}/a > 1/a$ we have

$$|D_x^\beta \partial\tilde{q}(x_1,x_2,0)/\partial\xi_2| \leqq C_\beta M^{\beta_1+1} a^{-\beta_2} B_2^{-1} \leqq C_\beta \rho M^{\beta_1+1} A_2 B_2^{\beta_2} \leqq$$
$$\leqq C_\beta N \rho M^{\beta_1+1} B_2^{\beta_2}/|\xi_2|$$

which gives a bound of the form (4.28) for $\xi_2\,\partial\tilde{q}(x_1,x_2,0)/\partial\xi_2$. If we write $\tilde{M}(x,\xi') = M(x_1,\chi_a(x',\xi'))$, it follows in view of Lemma 4.1, where we take $A_1 = 1/M\rho$ and $K = M$, that

$$(4.29)\quad \tilde{M}(x,\xi') \leqq C_N M \text{ if } |x_1M|<1,\ |x_2B_2|<1,\ |\xi_2A_2| \leqq N,\ |(x'',\xi'')| < \sqrt{\rho M}.$$

When (4.18) is fulfilled but not (4.18)′ we get the same conclusion with B_2 replaced by $1/\sqrt{\rho M}$ and χ_a equal to the orthogonal symplectic transformation such that $\tilde{q}(x,\xi') = q(x_1,\chi_a(x',\xi'))$. We shall use this convention throughout in order to have uniform notation.

A standard neighborhood Ω_0 of 0 is now defined by

$$\Omega_0 = \{(x_1,\chi_a(x',\xi'));\ (x,\xi') \in \tilde{\Omega}_0\}$$

where

$$\tilde{\Omega}_0 = \{(x,\xi'); |x_1 M| < 1, |x_2| < 1/B'_2, |\xi_2| < R_2, |(x'',\xi'')| < R_2\}$$

with the notation $1/B'_2 = \max(1/B_2, R_2)$ and $R_2 = \lambda^{-\kappa_2}$,

$$\kappa < \kappa_2 < 1/(k+1) .$$

This guarantees that $R_2\varepsilon$ is large and that $R_2^2/M\rho$ is small for small λ. We define $\Phi_0 \in C_0^\infty(\Omega_0)$ so that $\tilde{\Phi}_0(x,\xi') = \Phi_0(x_1, \chi_a(x',\xi'))$ is given by

$$\tilde{\Phi}_0(x,\xi') = \phi(Mx_1)\phi(B'_2 x_2)\,\phi(\xi_2/R_2)\phi((|x''|^2 + |\xi''|^2)/R_2^2)$$

where ϕ is the function used for the same purpose in case I.

(4.29) is only applicable in a small part of $\tilde{\Omega}_0$, since $1/A_2 < 1/\varepsilon << R_2$. We want to show now that $\tilde{M}$ is much larger than $M = \tilde{M}(0)$ when $A_2\xi_2$ is large. In fact, for $j \leqq k$ we have by (4.25) if (4.18)$'$ is valid

$$\rho\tilde{M}^{j+1}(x,\xi') \geqq |\tilde{q}^{(j)}(x,\xi')| \geqq |\xi_2\, \partial\tilde{q}^{(j)}(x_1,x_2,0)/\partial\xi_2| - C\rho M^{j+1}$$

so

$$\max_{j \leqq k} |(\xi_2/\rho)\partial\tilde{q}^{(j)}(x_1,x_2,0)/\partial\xi_2|^{1/(j+1)} \leqq \tilde{M}(x,\xi') + CM . \tag{4.30}$$

We know that there is some $j \leqq k/2$ such that

$$|\partial\tilde{q}^{(j)}(x_1,x_2,0)/\partial\xi_2|/\rho \geqq c\, M^{j+1}/\rho B_2 = c\, M^{j+1} A_2 ,$$

so the left hand side of (4.30) is bounded from below by

$$cM|\xi_2 A_2|^{2/(k+2)}$$

if $|\xi_2 A_2| > 1$. This allows us to remove CM from the right hand side of (4.30), and we obtain for a suitable N

$$\begin{aligned} cM|\xi_2 A_2|^{2/(k+2)} &\leqq \max_{j \leqq k} |(\xi_2/\rho)\partial\tilde{q}^{(j)}(x_1,x_2,0)/\partial\xi_2|^{1/(j+1)} \\ &\leqq \tilde{M}(x,\xi') \text{ if } |\xi_2 A_2| > N . \end{aligned} \tag{4.31}$$

(When (4.18) is valid but not (4.18)′ we may put $x_2 = 0$ in the middle member of this inequality; it follows from (4.9) then.)

From (4.31) it follows easily that all points in Ω_0 corresponding to points in $\tilde{\Omega}_0$ with $|\xi_2/R_2| > c$ for some fixed $c > 0$ belong to case I if λ is small. In fact, by (4.23) and (4.31) we have

$$|d_{x'\xi'}\tilde{q}^{(j)}(x,\xi')| \leqq |d_{x'\xi'}\tilde{q}^{(j)}(x_1,x_2,0)| + C\,M^j R_2$$
$$\leqq \tilde{M}(x,\xi')^{j+1}\rho/|\xi_2| + C\,M^j\sqrt{\rho M} << \varepsilon\rho\tilde{M}(x,\xi')^{j+1}$$

if $|\xi_2| >> 1/\varepsilon$. Since the first order derivatives of χ_a^{-1} are bounded uniformly the same conclusion holds for the differential of $q^{(j)}$. From the results proved in case I it therefore follows that

$$\tilde{M}(x,\xi') \leqq C \max_{j\leqq k} |(\xi_2/\rho)\partial\tilde{q}^{(j)}(x_1,x_2,0)/\partial\xi_2|^{1/(j+1)}, \text{ if } (x,\xi')\in\tilde{\Omega}_0 \text{ and } |\xi_2| > cR_2 . \tag{4.32}$$

By (4.29) with $N = 0$ and (4.31) we know that

$$\{(x,\xi')\in\tilde{\Omega}_0;\ \tilde{M}(x,\xi')\leqq C_0 M\}$$

contains $\{(x,\xi')\in\tilde{\Omega}_0;\ \xi_2 = 0\}$ and is contained in

$$\{(x,\xi')\in\tilde{\Omega}_0;\ |\xi_2 A_2| < C'\}$$

for some constant C'. Hence

$$\{(x,\xi')\in\Omega_0;\ M(x,\xi')\leqq C_0 M\} \tag{4.33}$$

contains a hypersurface with normal changing arbitrarily little for small λ, because R_2/a and $1/B_2 a$ are very small then, and the set (4.33) is contained in the $2C'/A_2$ neighborhood of the surface. Since $1/A_2 < 1/\varepsilon << R_2$ this means that (4.33) is very close to a hyperplane.

Let $(0, y', \eta')$ be a point in $\{(x',\xi');\ |x'|^2 + |\xi'|^2 \leqq (R_2/2)^2\}$ where $\tilde{M}(0,x',\xi')$ is minimal. By (4.31) we know then that $|\eta_2 A_2|$ has a fixed

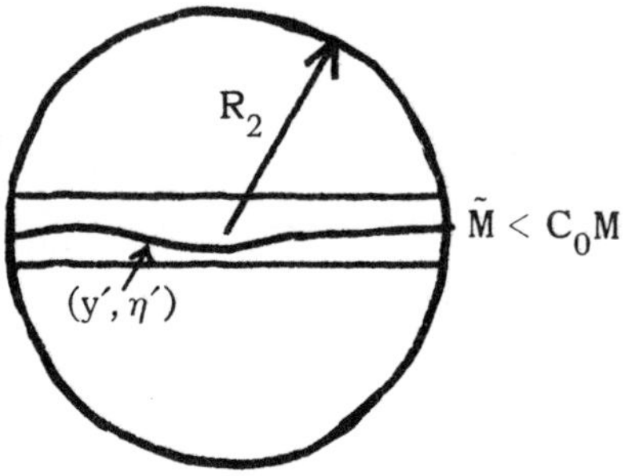

bound, for the minimum is $\leqq M$. Through (y', η') there passes, by the preceding argument, a hypersurface on which $\tilde{M}(0, x', \xi') \leqq C_0\tilde{M}(0, y', \eta') < C_0 M$, such that the normal varies very little for small λ. It is clear that it must be close to the ξ_2 axis then so the surface must intersect the ξ_2 axis at a point $(0, 0, \eta''_2, 0)$ at distance $\leqq C/A_2 << R_2$ from 0. Then we know that

$$\tilde{M}(0, 0, \eta''_2, 0) \leqq C_0\tilde{M}(0, x', \xi')$$

for all (x', ξ') at distance $< R_2/2$ from 0. We shall now analyze the behavior of the function M near $(0, 0, \eta''_2, 0)$. In doing so it is convenient to place the origin there and define the canonical change of coordinates relative to this point. Thus we assume now that for a fixed C_0

$$\text{(4.34)} \qquad M = \tilde{M}(0) \leqq C_0\tilde{M}(0, x', \xi') \text{ when } |x'|^2 + |\xi'|^2 < (R_2/3)^2 .$$

We shall see in a moment that such an estimate must then be valid for some other C in all of $\tilde{\Omega}_0$.

It follows from (4.25) that for fixed N

$$R(x, \xi') = \tilde{q}(x_1/M, x_2/B_2, \xi_2 B_2, x'', \xi'')/M - F(x_1, x_2) - G(x_1, x_2)\xi_2$$

is very small for $|x_1| < 1$, $|x_2| < 1$, $|\xi_2| < \rho N$, $|(x'', \xi'')| < R_2$ if

$$F(x_1, x_2) = \tilde{q}(x_1/M, x_2/B_2, 0)/M$$

$$G(x_1, x_2) = B_2/M\, \partial\tilde{q}(x_1/M, x_2/B_2, 0)/\partial\xi_2 .$$

In fact, (4.25) shows that R can be estimated by a positive power of λ, and since derivatives of high order are very small, this is true for the derivatives also. It follows that the difference between

$\tilde{M}(x_1/M, x_2/B_2, \xi_2 B_2, x'', \xi'')/M$ and the "M function" corresponding to $\xi_1 + i(F(x_1, x_2) + G(x_1, x_2)\xi_2)$ can be estimated by a positive power of λ for the values of the arguments in question.

As already observed, it follows from (4.21) and (4.23)$'$ that F/ρ for small λ is very close in C^∞ to a polynomial of degree k with fixed bounds for the coefficients, and we have also seen (cf. (4.26), (4.27)) that G is very close to a normalized polynomial in x_1 only, of degree $\leqq k/2$. The following lemma will give information on the commutators which will also be useful in section 5.

LEMMA 4.2. *Let* m_1 *and* m_2 *be fixed integers and let* $F_0(x_1, x_2), G_0(x_1)$ *be real valued polynomials of degree* $\leqq m_j$ *with respect to* x_j. *Set* $L_1 = \xi_1$, $L_2 = F_0(x) + G_0(x_1)\xi_2$ *and define* $L_I = \{L_{i_1}, \{L_{i_2}, \cdots\}\}$. *Then* $L_I = 0$ *when* $|I| > m = (m_1+1)(m_2+1)$. *Let*

$$\min_{\xi} \max_{I} |L_I(0, \xi)/\rho| = 1 , \tag{4.35}$$

$$\max |G_0^{(j)}(0)| = 1 . \tag{4.36}$$

If $Q = G_0\, \partial F_0/\partial x_1 - F_0 \partial G_0/\partial x_1$ *it follows that*

$$\rho/C_1 \leqq \sum |D^\alpha Q(0)| \leqq C_2 \rho . \tag{4.37}$$

The upper bound alone implies that there is a polynomial $H_0(x_2)$ *of degree* $\leqq m_2$ *such that the coefficients of* $F_0(x_1, x_2) - G_0(x_1) H_0(x_2)$ *are bounded by* $C_3 \rho$. *The lower bound in* (4.37) *implies*

$$\max_{I} |L_I(x, \xi)/\rho| \geqq C_4^{-1}(1+|x|)^{-N} . \tag{4.38}$$

The constants $C_1, \cdots, C_4$, N *depend only on* m_1 *and* m_2.

Proof. The Poisson brackets $\{L_1, \{L_1, \cdots, \{L_1, L_2\}\}\}$ are of the form

$$F_0^{(j)} + G_0^{(j)}\xi_2 \tag{4.39}$$

where $F_0^{(j)} = \partial^j F_0/\partial x_1{}^j$. The Poisson bracket with L_2 is

$$G_0 \partial F_0^{(j)}/\partial x_2 - G_0^{(j)}\partial F_0/\partial x_2 \tag{4.40}$$

and after that Poisson bracket with L_2 is equivalent with application of $G_0\partial/\partial x_2$. If we assign the weights 1 and m_1+1 to x_1 and x_2, then the weight of F_0 is at most $m_2(m_1+1)+m_1 = m-1$, and the weight of (4.40) is at most $m-2-j$. Taking Poisson bracket with L_1 or L_2 decreases the weight so L_I is of weight at most $m-|I|$ which implies $L_I = 0$ when $|I| > m$.

Since

$$Q = G_0\{L_1, L_2\} - G'_0 L_2$$

and the Hamilton field of $H_{L_1}{}^j L_2 = \partial^j L_2/\partial x_1{}^j$ is $G_0^{(j)}\partial/\partial x_2 \mod \partial/\partial\xi$, we have

$$(G_0^{(j)})^{\beta_2}(\partial/\partial x)^\beta Q = ((\operatorname{ad} H_{L_1})^j H_{L_2})^{\beta_2} H_{L_1}{}^{\beta_1}(G_0 H_{L_1} L_2 - G'_0 L_2). \tag{4.41}$$

The right hand side is a linear combination of L_I with coefficients which are polynomials in x_1 with bounded coefficients. By (4.36) and Taylor's formula we have

$$\max |G_0^{(j)}(x_1)| \geq C(1+|x_1|)^{-m_1},$$

so (4.41) implies the estimate

$$\sum |D^\alpha Q(x)| \leq C(1+|x_1|)^N \max_I |L_I(x,\xi)|.$$

Hence the upper bound in (4.37) follows from (4.35), and the lower bound in (4.37) gives (4.38) since

$$\sum |D^\alpha Q(x)| \geq C(1+|x_1|)^{1-2m_1}(1+|x_2|)^{-m_2}\sum |D^\alpha Q(0)|.$$

The linear map

$$F_0 \to G_0 \partial F_0/\partial x_1 - F_0 \partial G_0/\partial x_1$$

from polynomials of degree $\leqq m_1$ in x_1 and $\leqq m_2$ in x_2 to polynomials of degree $\leqq 2m_1-1$ in x_1 and $\leqq m_2$ in x_2 has as kernel the set of all F_0 such that $F_0(x_1, x_2) = G_0(x_1) H_0(x_2)$ for some polynomial H_0 of degree $\leqq m_2$, given for example by $H_0(x_2) = F_0^{(j)}(0, x_2)/G_0^{(j)}(0)$ where $|G_0^{(j)}(0)| = 1$. The dimension of the kernel is independent of G_0 and since the set of G_0 satisfying (4.36) is compact, it follows that the inverse modulo the kernel has a fixed bound. Thus

$$F_0(x_1, x_2) = G_0(x_1) H_0(x_2) + F_1(x_1, x_2)$$

where the largest coefficient of F_1 is at most a constant times the largest coefficient in Q. Now $L_2(x, \xi) = F_1(x_1, x_2) + G_0(x_1)(\xi_2 + H_0(x_2))$ and since $\xi_2 + H_0(x_2)$ can be taken as a new canonical coordinate instead of ξ_2 the condition (4.35) does not change if F_0 is replaced by F_1 in L_2. Taking $\xi = 0$ in (4.35) and recalling from the beginning of the proof that $L_I(0, 0)$ is a linear combination with bounded coefficients of the coefficients of F_1, we conclude from (4.35) that there is a fixed lower bound for the largest coefficient of F_1/ρ. This gives the lower bound in (4.37) and completes the proof.

We shall apply Lemma 4.2 to polynomials F_0 and G_0 which differ from F and G by $\mathcal{O}(\lambda^c)$ for some $c > 0$. Since F_0/ρ has uniformly bounded coefficients we have

$$|F_0^{(j)}(0) + G_0^{(j)}(0)\xi_2| > \rho$$

if $|G_0^{(j)}(0)\xi_2| > C\rho$. Hence it follows from (4.34) that (4.35) is valid with ρ replaced by a constant times ρ, and (4.38) then shows that (4.34) is valid in $\tilde{\Omega}_0$ with another constant when $|\xi_2 A_2|$ has a fixed bound. The same is true for large $|\xi_2 A_2|$ by (4.31). We state the result as a lemma after the following

DEFINITION 4.3. We shall say that Ω_0 is an admissible neighborhood of type II if (4.13) is not fulfilled but (4.34) is.

LEMMA 4.4. *If* Ω_0 *is an admissible neighborhood of type* II *then*

$$M(0) \leq C\, M(x,\xi') \quad \textit{if} \quad (x,\xi') \in \Omega_0 . \tag{4.34$'$}$$

The estimates which we have proved in Ω_0 remain valid with some other constants in the double of Ω_0 or more generally the neighborhood obtained if B_2^{-1} and R_2 are multiplied by fixed constants in the definition. It follows that for every $(x,\xi') \in \Omega_0$ the minimum of $M(x_1, y', \eta')$ for $|y'-x'|^2 + |\eta'-\xi'|^2 < 2R_2^2$ is bounded from above and below by constants times $M(0)$ and is attained only at points mapped by χ_a^{-1} to the set where $|\xi_2 A_2| < N$ for some fixed N. This implies that if admissible neighborhoods of type II with different centers overlap, then there is a fixed bound for the quotient of the values of M at the centers, and one neighborhood is contained in the other increased by a fixed factor.

It is now a routine matter to construct an appropriate covering of the set

$$\Omega = \{(x,\xi');\ |x_1| < 1/2,\ |(x',\xi')| < 1/2\lambda\} .$$

(Recall that we started from (4.1) in a set twice as large.) First we choose a (necessarily finite) maximal sequence of admissible neighborhoods $\Omega_1, \Omega_2, \cdots$ of type II with centers in Ω and corresponding functions $\Phi_j \in C_0^\infty(\Omega_j)$ defined as Φ_0 above such that the center of Ω_k is not in the inner half of Ω_j if $k > j$. Since overlapping neighborhoods have quite similar shape and become disjoint if they are shrunk by a fixed factor, it is clear that there is a fixed upper bound for the number of Ω_j which have a point in common. For later use we also define a function $\psi_j \in C_0^\infty$ with support in the set where $\Phi_j = 1$ by writing for example

$$\psi_0 \circ \chi_a = \tilde{\Phi}_0(4x/3, 4\xi'/3)$$

when the center is at 0 and χ_a is the corresponding canonical transformation. Then $\psi_j = 1$ at all points at distance $< cR_2$ in the $x'\xi'$ variables from the inner half of Ω_j, because $9/16 > 1/2$. It follows that the set

$$\omega = \{(x,\xi') \in \Omega;\ \psi_j(x,\xi') < 1 \text{ for every } j\}$$

contains only points of type I, for every point of type II is at distance $\mathcal{O}(1/\varepsilon)$ in the $x'\xi'$ variables from the center of an admissible neighborhood of type II. We can therefore cover ω by standard neighborhoods ω_j of type I with corresponding $\phi_j \in C_0^\infty(\omega_j)$ chosen as above, so that

$$\omega \subset \{(x,\xi');\ \phi_j(x,\xi') = 1 \text{ for some } j\}.$$

If we write

$$\Psi_j(x,\xi') = \psi_j(x,\xi') \prod_{i<j} (1-\psi_i(x,\xi'))$$

then supp Ψ_j is contained in the set where $\Phi_j = 1$, and

$$\Psi(x,\xi') = \sum \Psi_j(x,\xi') = 1 - \prod (1-\psi_j(x,\xi'))$$

is equal to 1 in $\Omega \setminus \omega$. The support of $d\Psi$, intersected with Ω, is covered by the sets where $\Phi_j = 1$ as well as by the sets where $\phi_k = 1$, which will make it possible to use Ψ in section 6 to separate the study of case I from the study of case II.

5. *Estimates for the localized operators*

In section 4 we proved that the symbol of a subelliptic operator is locally very close to that of a first order differential operator in one or two variables. We shall now discuss estimates for such operators, starting with the case of ordinary differential operators. The results are of course well known then (see Treves [10]) but we shall use somewhat unusual proofs which prepare for those needed in the case of two variables. All norms are L^2 norms in this section, and k is a fixed integer ≥ 0.

LEMMA 5.1. *Let* G *be a real valued* C^∞ *function in* $I = (-1,1)$ *such that*

$$|G^{(k+1)}| \leqq 1 \ \ in \ \ I, \tag{5.1}$$

$$\max_{j \leqq k} |G^{(j)}(0)/\rho|^{1/(j+1)} = 1. \tag{5.2}$$

If ρ *is large it follows that*

$$\rho^{1/(k+1)} \|v\| \leqq C(\|Dv\| + \|Gv\|), \ v \in C_0^\infty(I). \tag{5.3}$$

Proof. We may replace G by its Taylor polynomial of degree k since the error term in Taylor's formula is bounded by (5.1). Thus assume that G is a polynomial of degree k and that $\|Dv\| + \|Gv\| = 1$. Then

$$\|\tau_h v - v\| \leqq |h|, \ \ \|e^{ihG}v - v\| \leqq |h|$$

if τ_h is the one parameter group of unitary operators generated by D, that is, $\tau_h v = v(\cdot + h)$. If U_j are arbitrary unitary operators then

$$\|U_1 \cdots U_N v - v\| \leqq \sum_1^N \|U_1 \cdots U_{j-1}(U_j v - v)\| \leqq \sum_1^N \|U_j v - v\|$$

so we obtain with $\Delta_h = \tau_h$ −identity

$$\|\exp(ih\Delta_h G)\, v - v\| = \|\tau_h e^{ihG} \tau_{-h} e^{-ihG} v - v\| \leqq 4|h|.$$

Repetition of the argument gives inductively

$$\|\exp(ih\Delta_h{}^j G)\, v - v\| \leqq (2^j \cdot 3 - 2)|h|.$$

Now there is a triangular matrix (c_{ij}) such that for polynomials of degree k

$$h^\ell (d/dx)^\ell G = \sum_{j \geqq \ell} c_{\ell j} \Delta_h{}^j G.$$

Hence

$$\|(\exp i\, h^{j+1} G^{(j)})v - v\| \leqq C|h| ,$$

which proves that

$$\|(\exp ih\, G^{(j)})v - v\| \leqq C|h|^{1/(k+1)}, \ |h| < 1 . \tag{5.4}$$

Now (5.2) implies by Taylor's formula

$$C_2\rho \leqq \sum_j |G^{(j)}(x)| \leqq C_1\rho, \ x \in I . \tag{5.2$'$}$$

If we choose h in (5.4) so that $hC_1\rho = 1$, then (5.3) follows since $|\exp ih\, G^{(j)}(x_1) - 1|$ is not small for all j.

LEMMA 5.2. *Assume in addition to the hypotheses in Lemma* 5.1 *that* G *does not change sign from* $+$ *to* $-$ *for increasing* x. *Then*

$$\rho^{2/(k+1)} \int_{-1/2}^{1/2} |v|^2\, dx \leqq C\left(\rho^2 \int_{-1}^{1} |Dv + iGv|^2\, dx + \int_{-1}^{1} |v|^2\, dx\right) \tag{5.5}$$

if $v \in C^\infty(I)$ *and* ρ *is large enough.*

Proof. The estimate (5.2)$'$ is valid not only for the Taylor polynomial of G but for G itself, so if δ is small it follows from the mean value theorem that the set where $G = \delta\rho$ is discrete. If $v \in C_0^\infty(I)$ and we write $f = Dv + iGv$, $g = G/\rho$, then

$$\int_{|g|>\delta} |f|^2/|g|\, dx = \int_{|g|>\delta} (|Dv|^2 + |\rho g v|^2)/|g|\, dx - \rho \int_{|g|>\delta} \operatorname{sgn} g\, d|v|^2 .$$

The last term here can be integrated. At the boundary of an interval where $|g| < \delta$ and g has changed sign from $-$ to $+$ we obtain two positive terms, which we may drop, and at the others we obtain the difference between $|v|^2$ at the end points. The sum of these terms is bounded by

$$2\rho \int_{|g|<\delta} |v\, Dv|\, dx \leq 1/\delta \int_{|g|<\delta} |Dv|^2\, dx + \delta\rho^2 \int_{|g|<\delta} |v|^2\, dx\,.$$

Adding the estimate

$$2/\delta \int_{|g|<\delta} |Dv|^2\, dx + \rho^2/\delta \int_{|g|<\delta} |gv|^2\, dx \leqq 1/\delta \int_{|g|<\delta} (3\,|f|^2 + 7\rho^2\delta^2|v|^2)\, dx$$

we obtain

$$\int (|Dv|^2 + \rho^2|g|^2|v|^2)/\max(|g|,\delta)\, dx \leqq 3 \int |f|^2/\max(|g|,\delta)\, dx + $$
$$+ 8\rho^2\delta \int |v|^2\, dx\,.$$

If we apply (5.3) and choose δ so that

$$\rho^2\delta = 1 \tag{5.6}$$

it follows when ρ is large that

$$\rho^{2/(k+1)} \int |v|^2\, dx \leqq C \int |f|^2/\max(|g|,\rho^{-2})\, dx\,. \tag{5.7}$$

Now we can choose a cutoff function $\phi \in C_0^\infty(I)$ such that $0 \leqq \phi \leqq 1$, $\phi = 1$ in $(-1/2, 1/2)$, $|\phi'| \leqq C$, and $|g| > 1/C$ in $\operatorname{supp} \phi'$, for g is very close to a normalized polynomial and has therefore a fixed lower bound on intervals of fixed length $\subset (\pm 1/2, \pm 1)$. Since

$$(D + i\rho g)(\phi v) = \phi(D + i\rho g)v + v\, D\phi$$

and $|g| > 1/C$ in $\operatorname{supp} D\phi$, we obtain (5.5) if (5.7) is applied to ϕv.

LEMMA 5.3. *Assume that* G *satisfies* (5.1) *in an interval* I *of length at least* 1, *that* G *has no sign change from* $+$ *to* $-$ *for increasing* x *and that*

$$M_1(x) = \max_{j \leqq k} |G^{(j)}(x)|^{1/(j+1)}$$

is large for every $x \in I$. *Then we have with* C *independent of* G, u *and* I

(5.8) $$\|M_1 u\| \leqq C \|(D + iG)u\|, \; u \in C_0^\infty(I) .$$

Proof. Let ρ be a positive number which will be chosen later on in the proof, and set

$$M_\rho(x) = \max_{j \leqq k} |G^{(j)}(x)/\rho|^{1/(j+1)} .$$

We assume that $M_1(x) > \rho$ for every $x \in I$, which implies $M_\rho(x) > 1$. Now consider

$$v(y) = u(x + y/M_\rho(x)), \quad h(y) = f(x + y/M_\rho(x))$$

where $f = (D + iG)u$. Then

$$(D + iG(x + y/M_\rho(x))/M_\rho(x))\, v(y) = h(y)/M_\rho(x)$$

and an application of Lemma 5.2 gives

$$\rho^{2/(k+1)} \int_{|x-y|M_\rho(x) < 1/2} |u(y)|^2 M_\rho(x)^3\, dy$$

$$\leqq C \int_{|x-y|M_\rho(x) < 1} (\rho^2 |f(y)|^2 + M_\rho(x)^2 |u(y)|^2) M_\rho(x)\, dy .$$

If we apply (5.2)′ to $G(x + y/M_\rho(x))/M_\rho(x)$ it follows that the ratio $M_\rho(x)/M_\rho(y)$ has fixed upper and lower bounds when $|x-y|M_\rho(x) < 1$, so we may replace $M_\rho(x)$ by $M_\rho(y)$ in the integrals. If C is large enough we decrease the left hand side by integrating for $|x-y|M_\rho(y) < C^{-1}$ and increase the right hand side by integrating for $|x-y|M_\rho(y) < C$. Hence integration with respect to x also gives

$$\rho^{2/(k+1)} \int |M_\rho u|^2\, dy \leqq C\left(\rho^2 \int |f|^2\, dy + \int |M_\rho u|^2\, dy\right). \tag{5.9}$$

Fix ρ now so large that $\rho^{2/(k+1)} > 2C$. Then (5.9) implies (5.8).

REMARK. From (5.8) it follows that the large factor ρ^2 in the right hand side of (5.9) can be omitted. However, the proof of Lemma 5.2 was not sufficiently precise to give this right away. For that reason we made it even somewhat cruder than necessary for this did not matter in the proof of Lemma 5.3.

As a first step toward *a priori* estimates for differential operators in two variables we shall now use Lemma 5.3 to prove a simple and well known subelliptic estimate.

LEMMA 5.4. *Let* G *be a non-negative function belonging to a fixed bounded subset of* $C^N(\Omega)$, $\Omega = \{x \in \mathbf{R}^2;\ |x_1| < 1,\ |x_2| < 1\}$ *for some fixed sufficiently large* N, *and assume that for every* $x \in \Omega$

$$\max_{j \leqq k} |D_1{}^j G(x)| \geqq 1\,, \tag{5.10}$$

where k *is also fixed. If* K *is a fixed compact subset of* Ω *it follows that there is a fixed constant* C *such that*

$$\begin{aligned} \|GD_2 v\| + \|D_1 v\| + \|\,|D_2|^{1/(k+1)} v\| &\leqq \\ \leqq C(\|(D_1 + iG(x)D_2)v\| + \|v\|),\ v \in C_0^\infty(K)\,. \end{aligned} \tag{5.11}$$

Proof. We can change G outside K so that (5.10) is fulfilled for all x_2 and $\partial G/\partial x_2 = 0$ for $|x_2| > 1$. Then we no longer need to restrict the support of v in the x_2 direction. First we show that for fixed (x_2, ξ_2)

(5.12) $$\|G\xi_2 v\| + \|D_1 v\| + \| |\xi_2|^{1/(k+1)} v\| \leqq C \|(D_1 + iG(x_1, x_2)\xi_2)v\|, \quad v \in C_0^\infty(-1,1),$$

provided that ξ_2 is large. To do so we introduce $t = x_1|\xi_2|^\varepsilon$ as a new variable and observe that $G(t|\xi_2|^{-\varepsilon}, x_2)|\xi_2|^{1-\varepsilon}$ satisfies the hypotheses of Lemma 5.3 in the interval $|t| < |\xi_2|^\varepsilon$ if $(k+2)\varepsilon = 1$ and ξ_2 is large. When we return to the original x_1 variable the estimate (5.12) follows. Now choose $\kappa \in (1-1/(k+1), 1)$ and a partition of unity

$$1 = \sum \psi_j(\xi_2)^2$$

such that no more than two supports overlap and for suitably chosen t_j in $\operatorname{supp} \psi_j$ we have

$$|D^\alpha \psi_j(\xi_2)| \leqq C_\alpha(1+|\xi_2|)^{-\alpha\kappa}; \ |\xi_2 - t_j| < C\, t_j^\kappa \ \text{ if } \xi_2 \in \operatorname{supp} \psi_j .$$

Writing $(D_1 + iGD_2)v = f$ we have

$$\psi_j(D_2)f = (D_1 + iGt_j)\psi_j(D_2)v + iG(D_2 - t_j)\psi_j(D_2)v + i[\psi_j(D_2), G]\, D_2 v .$$

Here $\{[\psi_j(D_2), G]\, D_2\}$ can be regarded as a pseudo-differential operator of order $1-\kappa$ with values in ℓ^2. We choose $\tilde\psi_j$ satisfying the same conditions as ψ_j (with different constants) so that $\tilde\psi_j = 1$ in $\operatorname{supp} \psi_j$, and write $\phi_j(\xi_2) = (\xi_2 - t_j)\tilde\psi_j(\xi_2)$. Then

$$G(D_2 - t_j)\psi_j(D_2)v = G\phi_j(D_2)\psi_j(D_2)v = \phi_j(D_2)\, G\psi_j(D_2)v + [G, \phi_j(D_2)]\psi_j(D_2)v .$$

$\phi_j(D_2)$ has norm $\leqq C\, t_j^\kappa$ and $\{[G, \phi_j(D_2)]\psi_j(D_2)\}$ can be regarded as a pseudo-differential operator of order 0 with values in ℓ^2. Writing $v_j = \psi_j(D_2)v$ we conclude that

$$\sum \|(D_1 + iGt_j)v_j\|^2 \leqq C\Big(\|f\|^2 + \| |D_2|^{1-\kappa} v\|^2 + \|v\|^2 + \sum \|t_j^\kappa G v_j\|^2\Big) .$$

Now we apply (5.12) with $\xi_2 = t_j$ and $v = v_j$ in the left hand side and cancel terms which then appear with larger constants there than in the right hand side. This gives

$$\sum (\|G\, t_j v_j\|^2 + \|D_1 v_j\|^2 + \|\, |t_j|^{1/(k+1)} v_j\|^2) \leq C(\|f\|^2 + \|v\|^2) ,$$

and (5.11) is an immediate consequence.

In the rest of this section we shall consider a first order differential operator in two variables of the form

$$D_1 + i(F(x) + G(x) D_2) .$$

At first we shall assume that F and G are polynomials of fixed degrees and prove an analogue of Lemma 5.1. As a preliminary step we now compute $\exp itL$ where $L = F(x) + G(x) D_2$ and F, G are real valued, G independent of x_2, which makes L formally self-adjoint. We require that

$$u(x, t) = (\exp itL)\, u(x)$$

is equal to $u(x)$ when $t = 0$ and that

$$\partial u/\partial t = (iF(x) + G(x_1)\partial/\partial x_2) u .$$

Integration gives

$$u(x, t) = u(x_1, x_2 + G(x_1) t) \exp i \int_0^t F(x_1, x_2 + G(x_1)(t-s))\, ds$$

which has support close to supp u if t is small. The L^2 norm is of course unchanged. We take this formula as definition of $\exp itL$.

As in the proof of Lemma 5.1 we shall be concerned with functions u such that for some fixed a with $0 < a < 1$

(5.13) $$\|\exp itL\, u - u\| \leq t^a,\ 0 \leq t \leq T .$$

We can give this condition a sometimes more useful form by introducing

$$U_t = t^{-1}\int_0^t \exp isL\, u\, ds .$$

Since

$$iL\, U_t = (\exp itL\, u - u)/t$$

we obtain using (5.13)

(5.14) $$\|LU_t\| \leqq t^{a-1},\ \|U_t - u\| \leqq t^a/(a+1),\ 0 < t < T .$$

Conversely, assume that somehow we have obtained U_t with this property. Write $u = \sum_0^\infty u_k$ where

$$u_0 = U_T,\ u_k = U_{T/2^k} - U_{2T/2^k} \text{ for } k > 0 .$$

Then we have

$$\|Lu_k\| \leqq 2(T2^{-k})^{a-1},\ \|u_k\| \leqq 2(T2^{1-k})^a/(a+1)$$

so

$$\|e^{itL} u_k - u_k\| \leqq 2(T2^{-k})^{a-1} t,\ \|e^{itL} u_k - u_k\| \leqq 4(T2^{1-k})^a/(a+1) .$$

When $t < T2^{-k}$ we use the first estimate and otherwise we use the second one. Summing gives (5.13) with just a constant depending on a on the right hand side, so we see that (5.13) and (5.14) are essentially equivalent. An immediate consequence is that (5.13) implies

$$\|\exp(itLH)u - u\| \leqq C_a t^a,\ 0 < t < T ,$$

if H is a function of x_1 with $|H| < 1$ near supp u. In fact, (5.14) remains valid if L is replaced by HL. If $G = 0$ we may allow H to depend on x_2 also.

LEMMA 5.5. *Let* m_1 *and* m_2 *be fixed integers and let* $F(x_1, x_2)$, $G(x_1)$ *be real valued polynomials of degree* $\leqq m_j$ *with respect to* x_j *such that with* $L_1 = \xi_1$ *and* $L_2 = F(x) + G(x_1)\xi_2$, $L_I = \{L_{i_1},\{L_{i_2},\cdots\}\}$

$$\min_{\xi}\max_{I} |L_I(0,\xi)/\rho|^{1/|I|} = 1 . \tag{5.15}$$

Also assume that the coefficients of F/ρ *have fixed bounds and that*

$$\max_{j} |G^{(j)}(0)| = 1 . \tag{5.16}$$

If $N = (m_2+1)(m_1+1)$ *we have* $L_I = 0$ *when* $|I| > N$, *and for every compact set* $K \subset \mathbf{R}^2$ *there is a constant* C_K *such that for large* ρ

$$\rho^{1/N}\|u\| \leqq C_K(\|D_1u\| + \|(F(x)+G(x_1)D_2)u\|),\ u \in C_0^\infty(K) . \tag{5.17}$$

Proof. That $L_I = 0$ when $|I| > N$ was proved in Lemma 4.2. To prove (5.17) we shall replace the elementary identities used in the proof of Lemma 5.1 by the Campbell-Hausdorff formula. First note that the commutators of the operators $iL_j(x,D)$ are of the form $iL_I(x,D)$. Since all L_I with $|I|$ large vanish, the Campbell-Hausdorff formula implies (see Hörmander [5, p. 162]) that one can represent $\exp it^{|I|}L_I$ in the sense of formal power series as a product $\prod$ of a finite number of factors of the form $\exp \pm itL_j$. In fact, $\prod = \exp it^{|I|}L_I$ exactly since $\prod \exp -it^{|I|}L_I$ is an operator of the form

$$u \to u(\Phi(x,t)) \exp i\Psi(x,t)$$

for some analytic Φ and Ψ with $\Psi(x,0) = 0$ and $\Phi(x,0) = x$. Since

$$u(\Phi(x,t)) \exp i\Psi(x,t) - u(x)$$

vanishes of infinite order when $t = 0$, if $u \in C_0^\infty$, we obtain $\Psi = 0$ by taking $u = 1$ on a large compact set and then $\Phi(x,t) = x$ by taking u as a coordinate there. If u is normalized by

$$\|D_1u\| + \|(F+GD_2)u\| = 1$$

we have $\|L_j u\| \leqq 1$ and so

$$\|\exp itL_j u - u\| \leqq |t|$$

which implies as in the proof of Lemma 5.1 that

$$(5.18) \qquad \|\exp(i\, t^{|I|} L_I)u - u\| \leqq C\,|t|\ ,$$

hence that

$$(5.18)' \qquad \|\exp(itL_I)u - u\| \leqq C\, t^{1/N},\ 0 < t < 1\ .$$

This is analogous to (5.4), and to complete the proof we just have to show that L_I can be replaced by suitable functions of x in (5.18)′.

As pointed out above, before the statement of the lemma, the inequality (5.18)′ remains valid with another C if L_I is multiplied by a polynomial in x_1. If L'_1 and L'_2 are two such operators not containing D_1 then their commutators of high order vanish since the degree in x_2 is strictly decreasing. Hence the Campbell-Hausdorff formula gives as above (see [5, p. 161]) that $\exp it(L'_1 + L'_2)$ can be written as a product of a finite number of factors $\exp \pm itL'_j$. This is still true for a sum of more than two L'_j. If we write $Q = G\partial F/\partial x_1 - F\partial G/\partial x_1$ as in Lemma 4.2 it follows now from (4.41) that

$$(5.19) \qquad \|\exp(it(G^{(j)})^{\beta_2}(\partial/\partial x)^{\beta} Q)u - u\| \leqq C\, t^{1/N},\ 0 < t < 1\ .$$

In view of (5.16) and (4.37) we now obtain (5.17) as in the proof of Lemma 5.1 by choosing t as a small multiple of $1/\rho$.

The following lemma is a first step towards an analogue of Lemma 5.2.

LEMMA 5.6. *Assume in addition to the hypotheses in Lemma 5.5 that* G *is non-negative and* F/G *almost increasing in the sense that*

$$(5.20) \qquad \partial(F(x_1, x_2)/G(x_1))/\partial x_1 > -1 \ \textit{when}\ G(x_1) \geqq \rho^{-2},\ |x_1| < 1,\ |x_2| < 1,$$

$$(5.21) \qquad F(x_1, x_2)/G(x_1) - F(y_1, x_2)/G(y_1) > -1 \ \ \textit{if} \ \ -1 < y_1 < x_1 < 1, \quad |x_2| < 1 \ \ \textit{and} \ \ G(x_1) \geqq \rho^{-2},\ G(y_1) \geqq \rho^{-2}.$$

If ρ *is large enough and* $K = \{x;\ |x_1| \leqq 8/9,\ |x_2| \leqq 1\}$ *we have then*

$$(5.22) \quad \rho^{2/N} \int |v|^2 dx + \int |D_1 v|^2 dx \leqq C\Big(\int |(L_1 + iL_2)v|^2/\max(G, \rho^{-2})\, dx + \rho^{-2(1+1/m_1)} \int |D_1 D_2 v|^2 dx\Big),\quad v \in C_0^\infty(K).$$

Proof. We shall follow the proof of Lemma 5.2 closely. Thus we first observe that with $\delta = \rho^{-2}$ as in (5.6) and $H = F/G$ we have

$$(5.23) \quad \int_{G>\delta} (|L_1 v|^2 + |L_2 v|^2)/G\, dx = \int_{G>\delta} |(L_1 + iL_2)v|^2/G\, dx + E_1 + E_2$$

where

$$(5.24) \quad E_1 = 2\,\mathrm{Re} \int_{G>\delta} \partial v/\partial x_1\, \overline{D_2 v}\, dx = -2\,\mathrm{Re} \int_{G<\delta} \partial v/\partial x_1\, \overline{D_2 v}\, dx,$$

$$(5.25) \quad E_2 = 2\,\mathrm{Re} \int_{G>\delta} \partial v/\partial x_1\, \overline{Hv}\, dx = -\int_{G>\delta} |v|^2 \partial H/\partial x_1 dx - \int_{G=\delta} |v|^2 H\, dx_2 .$$

Here we have considered the set $G = \delta$ as the boundary of the set where $G < \delta$. If $a < x_1 < b$ is a component of this set, we have to estimate

$$\int |v(a, x_2)|^2 H(a, x_2)\, dx_2 - \int |v(b, x_2)|^2 H(b, x_2)\, dx_2 \leqq$$

$$\leqq \int |v(a, x_2)|^2 dx_2 + \int (|v(a, x_2)|^2 - |v(b, x_2)|^2) H(b, x_2)\, dx_2$$

where we have used (5.21). The normalization of G gives that we can find some $x_1 \in (\pm 8/9, \pm 1)$ where G is larger than a fixed constant, hence $|F(x_1)/G(x_1)| < C\rho$, so the monotonicity (5.21) shows that

$$|H(x)| < C\rho \quad \text{when} \quad x \in K \quad \text{and} \quad G(x_1) \geqq \delta\ . \tag{5.26}$$

Hence the second term can be estimated by

$$C\rho \int\limits_{a<x_1<b} |v\, D_1 v|\, dx\ .$$

Summing up, we obtain since there are at most m_1 intervals (a, b) to consider and the first term in E_2 can be estimated by (5.20)

$$\int\limits_{G>\delta} (|L_1 v|^2 + |L_2 v|^2)/G\, dx \leqq \int\limits_{G>\delta} |(L_1 + iL_2)v|^2/G\, dx \tag{5.27}$$

$$+ 2 \int\limits_{G<\delta} |D_1 v|\, |D_2 v|\, dx + C\Bigg(\rho \int\limits_{G<\delta} |vD_1 v|\, dx + \int\limits_{G>\delta} |v|^2 dx + \max_{G=\delta} \int |v|^2 dx_2\Bigg)\ .$$

In an interval where $|G| < 2\delta$ we have $|F| < 2C\delta\rho$ when $\delta < G < 2\delta$, in view of (5.26). Changing scales so that the interval becomes $(0, 1)$ for example one concludes immediately, since F and G are polynomials of fixed degrees, that $|F| < C_1\delta\rho$ when $|G| < 2\delta$. Hence

$$\int\limits_{G<\delta} (|L_1 v|^2 + |L_2 v|^2)\, dx \leqq 2 \int\limits_{G<\delta} |(L_1 + iL_2)v|^2 dx + 3 \int\limits_{G<\delta} |L_2 v|^2 dx \tag{5.28}$$

$$\leqq 2 \int\limits_{G<\delta} |(L_1 + iL_2)v|^2 dx + C\delta^2 \int\limits_{G<\delta} (|D_2 v|^2 + \rho^2 |v|^2)\, dx\ .$$

We divide by δ and add (5.27) where the integrals of $v\, D_1 v$ and $|D_1 v|\, |D_2 v|$ are estimated by means of the Cauchy-Schwarz inequality.

This gives

$$(5.29)\quad \int (|L_1 v|^2 + |L_2 v|^2)/\max(G,\delta)\, dx \leqq 2\int |(L_1 + iL_2)v|^2/\max(G,\delta)\, dx +$$

$$+ \int_{G<\delta} (|D_1 v|^2/2\delta + C\delta |D_2 v|^2)\, dx + C\int |v|^2 dx + C \max_{G=\delta} \int |v|^2 dx_2 \; .$$

The integral of $|D_1 v|^2/2\delta$ on the right hand side can be cancelled against half of the left hand side, and an application of Lemma 5.5 then gives

$$(5.30)\quad \rho^{2/N}\int |v|^2 dx + \int |D_1 v|^2 dx \leqq C\int |(L_1 + iL_2)v|^2/\max(G,\delta)\, dx +$$

$$+ C\delta \int_{G<\delta} |D_2 v|^2 dx + C\int |v|^2 dx + C \max_{G=\delta} \int |v|^2 dx_2 \; .$$

As already observed

$$|v(x)|^2 \leqq 2\int |v(t,x_2) D_1 v(t,x_2)|\, dt \leqq \int (\rho^{1/N}|v(t,x_2)|^2 + \rho^{-1/N}|D_1 v(t,x_2)|^2)\, dt \; .$$

If we integrate with respect to x_2 we see that the last term in (5.30) is much smaller than the left hand side for large ρ. The measure of the set of $x_1 \in (-1,1)$ with $G(x_1) < \delta$ can be estimated by $C\delta^{1/m_1}$ since it consists of at most m_1 intervals of length $< C\delta^{1/m_1}$, and

$$\int |D_2 v|^2 dx_2 \leqq C\int |D_1 D_2 v|^2 dx$$

for every x_1. If we use these estimates in (5.30), the estimate (5.22) is obtained.

We shall now prove a full analogue of Lemma 5.2. The statement is lengthy and the hypotheses made may look artificial until one sees how they occur in the proof of Lemma 5.8 which again will have a simple statement. We no longer assume that F and G are polynomials. To be able to return to the parameter ρ in Lemma 5.8 we replace it by t now.

LEMMA 5.7. *Assume that* F *and* G *are real valued* C^∞ *functions,* $G \geqq 0$, *in* $\Omega = \{x \in R^2;\ |x_j| < 1,\ j = 1, 2\}$ *and that for fixed* m_1, m_2 *and with* $N = (m_1+1)(m_2+1)$, $L_1 = \xi_1$ *and* $L_2 = F(x) + G(x)\xi_2$ *we have, with* ε *and* $1/t$ *small,*

$$\min_{\xi} \max_{|I| \leqq N} |L_I(0,\xi)/t|^{1/|I|} = 1, \tag{5.15$'$}$$

$$\max_{j \leqq m_1} |D_1^{\,j} G(0)| = 1, \tag{5.16$'$}$$

$$|D^\alpha F| \leqq t^{-4} \ \textit{if}\ |\alpha| \leqq N \ \textit{and}\ \alpha_1 > m_1 \ \textit{or}\ \alpha_2 > m_2, \tag{5.31}$$

$$|D^\alpha G| < \varepsilon < t^{-8} \ \textit{if}\ |\alpha| \leqq N \ \textit{and}\ \alpha_1 > m_1 \ \textit{or}\ \alpha_2 > 0, \tag{5.32}$$

$$|D^\alpha F| < 1/\varepsilon t^4 \ \textit{if}\ |\alpha| \leqq N, \tag{5.33}$$

$$\partial(F(x_1,x_2)/G(x_1,x_2))/\partial x_1 > -1/2 \ \textit{when}\ G(x_1) > 1/2t^2, \tag{5.20$'$}$$

$$F(x_1,x_2)/G(x_1,x_2) - F(y_1,x_2)/G(y_1,x_2) > -1/2 \ \textit{if}\ -1 < y_1 < x_1 < 1,\ |x_2| < 1,\ \min(G(x_1,x_2), G(y_1,x_2)) > 1/2t^2. \tag{5.21$'$}$$

Also assume that G *is in a fixed bounded subset of* $C^{N'}(\Omega)$ *for some* N' *which makes Lemma* 5.4 *applicable. If* K *is a compact subset of* $8\Omega/9$ *it follows for large* t *that*

$$\|D_1 v\|^2 + t^{2/N}\|v\|^2 \leqq C \int |(L_1 + iL_2)v|^2 \max(G, t^{-2})\,dx,\quad v \in C_0^\infty(K). \tag{5.34}$$

Proof. Let $F_0(x_1,x_2)$ and $G_0(x_1)$ be the part of the Taylor expansion of F and G at 0 with order $\leqq m_1$ in x_1 and $\leqq m_2$ resp. 0 in x_2. Then

G_0 satisfies (5.16) and we shall prove that (5.15) is fulfilled with $\mathcal{O}(1/t)$ added on the right hand side if L_I are replaced by the commutators L_I^0 formed from $L_1^0 = \xi_1$ and $L_2^0 = F_0(x) + G_0(x)\xi_2$. By induction one finds immediately that all L_I except L_1 are of the form

$$A(x_1, x_2) + B(x_1, x_2)\xi_2$$

where B is a polynomial in G and its derivatives while A is linear in F and its derivatives. When $|I| \leqq N$ only derivatives of order $< N$ occur, so (5.31)-(5.33) show that

$$|L_I(0,0) - L_I^0(0,0)| < C/t^4 .$$

If the minimum in (5.15)′ is attained at $(0, \xi_2)$ we have

$$|F^{(j)}(0) + G^{(j)}(0)\xi_2| < t$$

where j can be chosen so that $|G^{(j)}(0)| = 1$. Hence $|\xi_2| < 1/\varepsilon t^4$ by (5.33). We can reduce ξ_2 to 0 be a translation of the ξ_2 variable which only leads to a constant factor on the right hand side of (5.33). It follows then from the preceding remarks that

$$\min_{\xi} \max_{I} |L_I^0(0,\xi)/t|^{1/|I|} < 1 + C/t^4 .$$

By Lemma 4.2 it follows that there is a polynomial $H_0(x_2)$ of degree $\leqq m_2$ such that the coefficients of $F_0(x) - G_0(x_1)H_0(x_2)$ are bounded by a constant times t. If we replace v by $v \exp \int -i H_0(x_2)\,dx_2$ in (5.34) then F is replaced by $F - GH_0$ which does not affect (5.31) by more than a constant factor since the coefficients of H_0 are bounded by a constant times $1/\varepsilon t^4$. Obviously (5.20)′, (5.21)′ remain valid, and (5.33) is now replaced by a bound for the derivatives of F/t. Since L_2 has just been modified by introduction of $\xi_2 + H_0(x_2)$ as a new canonical variable instead of ξ_2, condition (5.15)′ is unchanged also.

From now on we may therefore assume that we have a fixed bound for the derivatives of F/t instead of just (5.33). Lemma 5.6 is then applicable to F_0, G_0, for the bounds (5.31), (5.32) have been chosen so small that (5.20), (5.21) for F_0, G_0 follow from (5.20)′, (5.21)′ when t is large. The elementary verification is left for the reader. Lemma 5.4 is also available and we shall prove (5.34) by splitting v into two parts corresponding to low and high frequency with respect to x_2.

Choose a function $\psi \in C_0^\infty(-8/9, 8/9)$ such that $\psi(x_2) = 1$ in a neighborhood of K, and let $\chi \in C_0^\infty(\mathbf{R})$ be equal to 1 in $(-1, 1)$. With a fixed κ satisfying $0 < \kappa < 1/N$ we set

$$\psi_t^{(1)}(x_2, \xi_2) = \psi(x_2)(1 - \chi(\xi_2/t^{1+\kappa})), \psi_t^{(2)}(x_2, \xi_2) = \psi(x_2)\chi(\xi_2/t^{1+\kappa}) .$$

We shall apply Lemma 5.4 to $\psi_t^{(1)}(x_2, D_2)v$ and Lemma 5.6 to $\psi_t^{(2)}(x_2, D_2)v$ but first we shall prove that

$$\int |(L_1 + iL_2)\psi_t^{(2)}(x_2, D_2)v|^2/\max(G, t^{-2})\,dx \leqq \tag{5.35}$$

$$\leqq 2\int |(L_1 + iL_2)v|^2/\max(G, t^{-2})\,dx + C\|v\|^2, \quad v \in C_0^\infty(K) .$$

This means that we must estimate the commutator $[L_2, \psi_t^{(2)}(x_2, D_2)]$. To do so we use the Beals-Fefferman calculus of pseudo-differential operators with weights $\phi = 1$ and $\Phi = t^{1+\kappa} + |\xi_2|$. A symbol a is called of weight $m(\xi)$ then if

$$|D_\xi^\alpha D_x^\beta a(x, \xi)| \leqq C_{\alpha\beta}\, m(\xi)\Phi(\xi)^{-|\alpha|} .$$

(What we need is actually a very small addition to the standard calculus of operators of type 1,0.) Since L_2 is a symbol of weight Φ and $\psi_t^{(2)}$ is of weight 1, the error committed if the commutator is replaced by the Poisson bracket has a norm $\mathcal{O}(t^{-1-\kappa})$. A term in the Poisson bracket

$$\{L_2(x, \xi_2), \psi_t^{(2)}(x_2, \xi_2)\}$$

where $\psi(x_2)$ is differentiated corresponds to an operator which becomes of weight $1/\Phi$ if this derivative is moved next to v, for it annihilates v. Thus we need only examine the symbol

$$\psi(x_2)\{L_2(x,\xi_2),\chi(\xi_2/t^{1+\kappa})\} = -t^{-1-\kappa}\psi(x_2)\partial L_2/\partial x_2\,\chi'(\xi_2/t^{1+\kappa})\,.$$

Since $|\partial G/\partial x_2| < 1/t$ the operator

$$-t^{-1-\kappa}\partial G/\partial x_2\, D_2\chi'(D_2/t^{1+\kappa})$$

has norm $\mathcal{O}(1/t)$, so what remains is to study

$$t^{-1-\kappa}\psi(x_2)\partial F/\partial x_2\chi'(\xi_2/t^{1+\kappa})\,.$$

In (5.35) it does not matter if we replace $G(x_1,x_2)$ by $G_0(x_1)$ since the difference is $\mathcal{O}(t^{-8})$. When $|G_0(x_1)| > 1/t^2$ we have $|F_0(x_1,x_2)| < Ct|G_0(x_1)|$ and when $|G_0(x_1)| < 1/t^2$ we have $|F_0(x_1,x_2)| < C/t$ as observed in the proof of Lemma 5.6. Since F_0 is a polynomial of fixed degree this is also true for the derivative with respect to x_2 and we conclude that

$$|\partial F_0(x)/\partial x_2|/t < C(G(x)+1/t)\,.$$

Since $(G+1/t)^2/\max(G,1/t^2)$ is bounded we obtain (5.35).

When we apply (5.22) to $\psi_t^{(2)}(x_2,D_2)v$ with F, G replaced by F_0, G_0 we observe that by (5.31) and (5.32)

$$\|(F-F_0+(G-G_0)D_2)\psi_t^{(2)}(x_2,D_2)v\| \leqq C\|v\|/t^4\,.$$

The last term on the right hand side of (5.22) becomes the square of

$$t^{-1-1/m_1}\|D_1D_2\psi_t^{(2)}(x_2,D_2)v\| \leqq C\,t^{\kappa-1/m_1}\|D_1v\|\,.$$

Since $\kappa < 1/m_1$ and $\|D_1v\|$ occurs in the left hand side of 5.34) this can be cancelled for large t.

When (5.11) is applied to $\psi_t^{(1)}(x_2, D_2)v = v_1$, we obtain

(5.36)
$$\|GD_2v_1\| + \|D_1v_1\| + \||D_2|^{1/(m_1+1)}v_1\| \leq$$
$$\leq C(\|(L_1+iL_2)v_1\| + \|v\| + \|Fv_1\|) .$$

From the calculus it follows at once that

$$t^{1/(m_1+1)}\|v_1\| \leq C(\||D_2|^{1/(m_1+1)}v_1\| + \|v\|) .$$

To estimate the last term on the right hand side of (5.36) we let h denote a C^∞ function on $\mathbf{R}$ which is 0 near 0 and $1/\xi_2$ near supp $1-\chi$, and we write $h_t(\xi_2) = h(\xi_2/t^{1+\kappa})$. Then

$$\psi_t^{(1)}(x_2, D_2) - t^{-1-\kappa}\psi(x_2)h_t(D_2)D_2\psi_t^{(1)}(x_2, D_2)$$

is of weight $1/\Phi$ near supp v, with the notation used in the proof of (5.35). Hence the norm is $< C\,t^{-1-\kappa}$. Since $|F| < C(G_0t+1)$ we obtain by separating the cases where $G_0 < 1/t$ and $G_0 \geq 1/t$

$$\|F v_1\| \leq C(t^{-\kappa}\|GD_2v_1\| + \|v\|) .$$

The first term on the right occurs with a larger coefficient on the left hand side of (5.36) so it can be cancelled. Summing up, we have proved

$$\|D_1v\|^2 + t^{2/N}\|v\|^2 \leq C\left(\int |(L_1+iL_2)v|^2/\max(G, t^{-2})\,dx + \|v\|^2\right)$$

which gives (5.34) when t is large.

Here is our final lemma, which is analogous to Lemma 5.3 in statement and in proof. As usual k is a fixed positive integer.

LEMMA 5.8. *Assume that* F *and* G *are real valued functions,* $G \geq 0$, *in* $\Omega = \{x \in \mathbf{R}^2;\ |x_1| < 1,\ |x_2| < 1\}$, *and that for some fixed* N', C

(5.37) $$|D^\alpha F| \leqq C\rho \quad when \quad |\alpha| \leqq N'$$

(5.38) $$|D^\alpha G| \leqq C/\rho \quad when \quad |\alpha| \leqq N' \quad and \quad \alpha_1 > k \quad or \quad \alpha_2 \neq 0 .$$

Also assume that with $L_1 = \xi_1$ *and* $L_2 = F + G\xi_2$

(5.39) $$\max_{|I| \leqq k+1} |L_I(x, \xi_2)/\rho|^{1/|I|} \geqq 1 \quad if \quad x \in \Omega,\ \xi_2 \in R ,$$

(5.40) $$\max_{j \leqq k} |\partial^j G(0)/\partial x_1{}^j| = 1 ,$$

(5.41) $$\partial(F/G)/\partial x_1 > -1 \quad when \quad G(x) > \rho^{-(k+1)} ,$$

(5.42) $$F(x_1, x_2)/G(x_1, x_2) - F(y_1, x_2)/G(y_1, x_2) > -1 \quad if$$
$$-1 < y_1 < x_1 < 1 \quad and \quad \min(G(x_1, x_2), G(y_1, x_2)) > \rho^{-k-1} .$$

If K *is a compact subset of* Ω *it follows for large* ρ *and* N' *that*

(5.43) $$\rho^{1/(k+1)} \|u\| + \|D_1 u\| \leqq C\|(D_1 + i(F(x) + G(x) D_2))u\|, \quad u \in C_0^\infty(K) .$$

Proof. The derivation of Lemma 5.8 from Lemma 5.7 is exactly parallel to that of Lemma 5.3 from Lemma 5.2 so we shall only sketch the main steps. Let $1 << t << \rho$; we shall fix t later on in the proof. We want to apply Lemma 5.7 with $m_1 = m_2 = k$ so we introduce

$$M(x) = \min_{\xi} \max_{|I| \leqq N} |L_I(x, \xi)/t|^{1/|I|}$$

$$1/B(x) = \max_{j \leqq k} |\partial^j G(x)/\partial x_1{}^j|/M(x)^{j+1} .$$

For large ρ it follows from (5.37) - (5.40) that

(5.44) $$(\rho/t)^{1/(k+1)} < M(x) < C\rho/t, \quad M(x)/3 < B(x) < 3M(x)^{k+1} .$$

We shall now change scales at x in the ratios $M(x)$, $B(x)$. To simplify notation we take $x = 0$ at first and write M and B instead of $M(0)$ and $B(0)$. After the scale change and division by M our operator has the symbol

$$\xi_1 + i(F(x_1/M, x_2/B)/M + G(x_1/M, x_2/B)\xi_2 B/M)\,. \tag{5.45}$$

The hypothesis (5.15)′ in Lemma 5.7 is valid for this operator by the definition of M, for symplectic dilations do not affect Poisson brackets, and (5.16)′ follows from the definition of B. Since

$$|D^\alpha F(x_1/M, x_2/B)/M| \leqq C_\alpha \rho M^{-\alpha_1-1} B^{-\alpha_2} \leqq C'_\alpha \rho M^{-|\alpha|-1}$$

we obtain (5.31) for $|\alpha| > k$ if t is fixed and ρ is large. (5.32) with $\varepsilon = 1/\rho$ follows from (5.38) if we observe that $B/M^{k+1} < 3$. The estimate (5.33) is obvious. If $GB/M > t^{-2}/2$ then

$$G > M/2t^2B > 1/6t^2M^k > \rho^{-k-1}$$

if ρ is large. Since $1/BM << 1$ and $1/B << 1$ the conditions (5.20)′, (5.21)′ follow from (5.41) and (5.42). Altogether this shows that (5.34) is applicable to the operator with symbol (5.45). As in the proof of Lemma 5.2 we modify (5.34) by cutting off an arbitrary v with a cutoff function ψ which is 1 for $|x_j| < 1/2$, $j = 1, 2$, and such that G is bounded from below in supp $\partial\psi/\partial x_1$. Then it follows that

$$\int_{|x_j|<1/2} (|D_1 v|^2 + t^{2/N}|v|^2)\,dx \leqq C \int_{|x_j|<1} (t^2|(D_1 + i(F(x_1/M, x_2/B)/M + G(x_1/M, x_2/B)\,B/M\,D_2)v|^2 + |v|^2)\,dx\,. \tag{5.34'}$$

Now observe that the proof of Lemma 5.7 also shows that M and B do not change by more than a fixed factor when $|x_1 M| < 1$ and $|x_2 B| < 1$. We can therefore return to the original variables in (5.34)′, multiply by

M^3B evaluated at a variable point y where we make the preceding construction instead of at the origin, and integrate with respect to y. Then we obtain

$$\|D_1 u\|^2 + t^{2/N}\|Mu\|^2 \leqq C(t^2\|(D_1 + i(F(x)+G(x)D_2))u\|^2 + \|Mu\|^2).$$

Now we fix t so large that the last term can be cancelled. In view of the first inequality in (5.44) we then obtain (5.43).

6. *Proof of the sufficiency in Theorem 3.4*

Let q be a function satisfying (4.1) in R^{2n-1} and (4.2), (4.3) when $|(x',\xi')| < 1/\lambda$, $|x_1| < 1$. Reversing the arguments which led from (3.2) to (3.3) and the scale changes at the end of section 3 one shows easily that the sufficiency in Theorem 3.4 follows if we show that for $h \in C_0^\infty$ with support in the ball of radius $\frac{1}{2}$

$$\lambda^{-2/(k+1)}\|h(\lambda x', \lambda D')u\| \leqq C(\|(D_{x_1} + iq(x, D'))u\| + \|u\|), \tag{6.1}$$

$$\text{if } u \in \mathcal{S} \text{ vanishes for } |x_1| > 1/2 .$$

The details of this standard argument are left for the reader. Set

$$U = h(\lambda x', \lambda D')u, \quad P = D_{x_1} + iq(x, D').$$

where u is assumed to be as in (6.1). Then

$$PU = h(\lambda x', \lambda D')Pu + [iq, h]u .$$

The calculus of general pseudo-differential operators shows that for the symbol of $[iq, h]$ there is an estimate of the form (4.1) but without the factor λ^{-2} in the right hand side, so it is bounded. Hence $PU = f$ where

$$\|f\| \leqq C(\|(D_{x_1} + iq(x, D'))u\| + \|u\|). \tag{6.2}$$

Next we split U by means of the function Ψ constructed at the end of section 4, thus $U = U_1 + U_2$ where

$$U_1 = (1-\Psi(x,D'))U, \quad U_2 = \Psi(x,D')U . \tag{6.3}$$

With $g = [P,\Psi(x,D')]U$ we have

$$PU_1 = (1-\Psi(x,D'))f - g, \quad PU_2 = \Psi(x,D')f + g . \tag{6.4}$$

To estimate g we observe first that

$$|D_{x'}^{\beta} D_{\xi'}^{\alpha} \Psi(x,\xi')| \leqq C_{\alpha\beta} R_2^{-|\alpha+\beta|} , \tag{6.5}$$

for $a >> R_2$ so such an estimate is preserved under composition with χ_a or its inverse, and in the definition of Ψ only a fixed number of terms can be non-zero at any point. From the calculus it follows therefore that $[iq,\Psi]$ is bounded apart from the term with symbol $\{q,\Psi(x,\xi')\}$. Now we have

$$D_{x_1}\Psi(x,\xi') + \{q,\Psi(x,\xi')\} = \sum r_j(x,\xi')M_j\Phi_j(x,\xi') \tag{6.6}$$

where M_j is the value of M at the center of Ω_j and

$$r_j(x,\xi') = M_j^{-1}(D_{x_1}+H_q)\Big(\psi_j(x,\xi')\prod_{i<j}(1-\psi_i(x,\xi'))\Big) \tag{6.7}$$

has support in the set where $\Phi_j = 1$. We shall prove an estimate of the form (6.5) for r_j. Admitting this for a moment we observe that the calculus gives a uniform bound for the norm of

$$[P,\Psi(x,D')] - \sum r_j(x,D')M_j\Phi_j(x,D')$$

for only the first term in the expansion of the symbol of the composition is non-zero and the remainder term can therefore be estimated by any desired power of λ. (Note also that the total number of terms is bounded by some power of $1/\lambda$.) Regarding $\{r_j\}$ as a vector valued pseudo-differential operator we may then conclude that

$$\|g\|^2 \leqq C\Big(\|U\|^2 + \sum \|M_j \Phi_j(x, D')U\|^2\Big). \tag{6.8}$$

We shall now supply the proof that r_j has a bound of the form (6.5). In the product (6.7) we need only consider factors such that the support of ψ_i overlaps that of ψ_j, thus only a fixed number. Since M_i/M_j is bounded for such indices i, it is clear that we just need to show such a bound for

$$r'_j(x, \xi') = M_j^{-1}(D_{x_1} + H_q)\psi_j(x, \xi').$$

In doing so we may consider the composition with χ_a; to conform with the notations in section 4 we assume $j = 0$ and set $\tilde{\psi} = \psi_0 \circ \chi_a$. In the new coordinates $r_{0'}$ is the sum of

$$r''_0(x, \xi') = M^{-1}\{\xi_2\, \partial\tilde{q}(x_1, x_2, 0)/\partial\xi_2, \tilde{\psi}(x, \xi')\},$$

the term $M^{-1} D_{x_1}\tilde{\psi}$ which obviously satisfies (6.5), and the Poisson bracket of $(\tilde{q}(x, \xi') - \xi_2\, \partial\tilde{q}(x_1, x_2, 0)/\partial\xi_2)/M$ and $\tilde{\psi}$. It satisfies (6.5) with an additional factor $\rho/R_2\sqrt{M\rho}$ on the right hand side since (4.25) is valid with B_2 replaced by $\min(B_2, 1/\sqrt{M\rho})$. Now

$$r''_0(x, \xi') = (M^s a^{-1} \partial\tilde{q}/\partial\xi_2)(B_2^{-1}\partial\tilde{\psi}/\partial x_2) + M^{-1}\, \xi_2\{\partial\tilde{q}/\partial\xi_2, \tilde{\psi}\}.$$

In the first term both factors have estimates of the form (6.5), and this is true for the second term even without the factor M^{-1} since $|\xi_2| < R_2$ and every differentiation of $\tilde{\psi}$ improves the estimate (6.5) by a factor $1/R_2$ while (4.23) gives good bounds for derivatives of $\tilde{q}$ of order $\geqq 2$. This completes the proof of (6.8).

We shall replace U by $U_1 + U_2$ in (6.8). If we prove that

$$\sum \|M_j \Phi_j(x, D')U_1\|^2 \leqq C\Big(\sum \|m_k \phi_k(x, D')U_1\|^2 + \|u\|^2\Big) \tag{6.9}$$

where m_k is the value of M at the center of ω_k, we shall then obtain

$$(6.10)\quad \|g\|^2 \leqq C\Big(\|u\|^2 + \sum_j \|M_j \Phi_j(x, D')U_2\|^2 + \sum \|m_k \phi_k(x, D')U_1\|^2\Big) .$$

To prove (6.9) it suffices to show that

$$(6.11)\qquad \Phi_j(x, D')U_1 = \sum a_{jk}(x, D')\phi_k(x, D')U_1 + S_j(x, D')u$$

where supp $a_{jk} \subset$ supp $\Phi_j \cap$ supp ϕ_k and a_{jk} satisfies (6.5) with R_2 replaced by R_1 while the norm of S_j is bounded by as high a power of λ as we please. To prove (6.11) we first observe that the construction of $\{\phi_k\}$ shows that

$$(6.12)\qquad 1 = \sum b_k(x, \xi')\phi_k(x, \xi') \text{ in } \Omega \cap \text{supp } (1-\Psi)$$

where b_k is of type R_1 in the sense that (6.5) is valid for b_k with R_2 replaced by R_1. The first approximation to (6.11) is then

$$a_{jk}{}^0(x, \xi') = \Phi_j(x, \xi')\, b_k(x, \xi')$$

which gives (6.11) apart from an error of the form $R_1^{-2}\Phi'_j(x, D')U_1$ where supp $\Phi'_j \subset$ supp Φ_j and Φ'_j is of type R_1. We iterate the construction with Φ_j replaced by Φ'_j and so on, which gives the desired form of (6.11) after a finite number of steps.

With (6.10) we have achieved a separation of the cases I and II of section 4, just as in the proof of Lemma 5.7 by means of the cutoff functions $\psi_t{}^{(1)}$ and $\psi_t{}^{(2)}$. We pass now to the study of case II; the similar and somewhat simpler arguments in case I will be sketched briefly later. The first step is to decompose U_2 in the pieces $\Phi_j U_2$. Since $\sum_j \Phi_j(x, \xi')^2 \geqq 1$ in supp Ψ we can write (with new b_j)

$$(6.13)\qquad 1 = \sum_j b_j(x, \xi')\Phi_j(x, \xi') \text{ in supp } \Psi$$

where b_j is of type R_2, that is, satisfies (6.5), and supp $b_j \subset$ supp Φ_j.

Just as in the proof of (6.11) we can then construct a_j with the same property and leading term b_j so that

$$U_2 = \sum a_j(x, D')\Phi_j(x, D')U_2 + Su$$

where $\lambda^{-2}S$ is of type R_2. Hence

$$\|U_2\|^2 \leqq C\Big(\sum \|\Phi_j(x, D')U_2\|^2 + \lambda^2\|u\|^2\Big). \tag{6.14}$$

Here we have of course used the calculus of vector valued pseudo-differential operators.

To gain control of $\Phi_j(x, D')U_2$ we must first estimate

$$f_j = P\Phi_j U_2 = \Phi_j PU_2 + [P, \Phi_j]U_2 .$$

The L^2 continuity of pseudo-differential operators gives

$$\sum \|\Phi_j PU_2\|^2 \leqq C\|PU_2\|^2 .$$

We can also write $[P, \Phi_j]$ as the sum of the operator with symbol $i^{-1}\{p, \Phi_j\}(x, D')$ and one of order 0 for which we also have L^2 continuity from scalar to ℓ^2 valued functions. By (6.13) we have in supp Ψ

$$\{p, \Phi_j\} = \sum a_{jk}{}^0 M_k \Phi_k$$

where

$$a_{jk}{}^0 = \{p, \Phi_j\} b_k / M_k .$$

Our discussion of the function r_j defined by (6.7) applies without change to show that $a_{jk}{}^0$ is of type R_2; again it is of course essential that M_j/M_k is bounded if supp $\Phi_j \cap$ supp $\Phi_k \neq \emptyset$. By the iterative argument used to establish (6.11) we can find a_{jk} of type R_2 with support in supp $\Phi_j \cap$ supp Φ_k so that

$$\{p, \Phi_j\}(x, D) U_2 = \sum a_{jk}(x, D') M_k \Phi_k(x, D') U_2 + S_j(x, D') u ,$$

where the norm of S_j is bounded by any desired power of λ. Summing up, we have therefore proved that

$$\sum \|f_j\|^2 \leqq C\left(\|PU_2\|^2 + \sum \|M_j \Phi_j U_2\|^2 + \|u\|^2\right). \tag{6.15}$$

In the equation $P\Phi_j U_2 = f_j$ we shall now replace P by an approximation. Let $j = 0$ at first to simplify notation, and write

$$r_0(x, \xi') = q(x, \xi') - L \circ \chi_a^{-1}(x, \xi')$$

where

$$L(x, \xi') = \tilde{q}(x_1, x_2, 0) + \xi_2 \partial \tilde{q}(x_1, x_2, 0)/\partial \xi_2 .$$

(A similar definition is made when $j \neq 0$.) We have assumed here that χ_a is globally defined which is clearly no restriction if χ_a is chosen of the special form discussed after (4.22). (However, $q \circ \chi_a = \tilde{q}$ has the bounds given in section 4 only in a part of the set where $|(x', \xi')| < ca$.) We want to estimate

$$\|r_0(x, D') \Phi_0(x, D') U_2\| .$$

Crude estimates show that for example

$$|D_{x'}^{\alpha} D_{\xi'}^{\beta} L \circ \chi_a^{-1}| < C\lambda^{-2} R_2^{-|\alpha+\beta|}$$

and this suffices to conclude from the calculus that the symbol of $r_0(x, D')\Phi_0(x, D')/\lambda^N$ for any N is of type R_2 outside supp Φ_0. To study r_0 near supp Φ_0 we form the composition with χ_a,

$$\tilde{r}_0(x, \xi') = r_0 \circ \chi_a(x, \xi') = \tilde{q}(x, \xi') - L(x, \xi') .$$

For $R(x, \xi') = \tilde{q}(x, \xi') - \xi_2 \partial \tilde{q}(x_1, x_2, 0)/\partial \xi_2$ we have the estimate

$$|D_{\xi'}^{\alpha} D_{x'}^{\beta} R(x, \xi')| \leq C_{\alpha\beta} M\rho(M\rho)^{-|\alpha+\beta|/2}$$

by (4.25) and (4.9). Taylor's formula gives

$$\tilde{r}_0(x, \xi') = R(x, \xi') - R(x_1, x_2, 0) = \sum_3^n T_j(x, \xi') x_j + \sum_2^n S_j(x, \xi') \xi_j$$

where

$$\sum_j |D_{\xi'}^{\alpha} D_{x'}^{\beta} T_j| + \sum_j |D_{\xi'}^{\alpha} D_{x'}^{\beta} S_j| < C_{\alpha\beta} M\rho(M\rho)^{-(1+|\alpha+\beta|)/2} .$$

In supp Φ_0 we have $|x''| + |\xi'| < R_2 << \sqrt{M/\rho}$ so $R_2 T_j/M$ and $R_2 S_j/M$ are symbols of type R_2. It follows that r_0/M is of type R_2 in Ω_0. The corresponding conclusion holds when $j \neq 0$. Hence

$$\sum \|r_j(x, D')\Phi_j(x, D')U_2\|^2 \leq C\Big(\sum \|M_j \Phi_j(x, D')U_2\|^2 + \|u\|^2\Big) ,$$

so (6.15) gives

(6.16) $$\sum \|(P - ir_j(x, D'))\Phi_j(x, D')U_2\|^2 \leq C\Big(\|PU_2\|^2 + \sum \|M_j \Phi_j(x, D')U_2\|^2 + \|u\|^2\Big) .$$

The symbol of $P - ir_0$ is $\xi_1 + iL \circ \chi_a^{-1}(x, \xi')$. Choosing χ_a in the special way indicated in section 4 it is easy to show, as we shall do later, that the corresponding operator is unitarily equivalent with $D_1 + iL(x, D')$, at least with a very small error. We shall therefore derive estimates for the inverse of this operator now.

We shall apply Lemma 7.4 to

$$F(x_1, x_2, \xi_2) = \tilde{q}(x_1/M, x_2/B_2, 0, \xi_2\sqrt{\rho M}, 0)/\rho M$$

where $|x_1| < 2$, $|x_2| < 2$, $|\xi_2| < 2$ say. ($x_1/2$ plays the role of t and a multiple of ξ_2 plays the role of y in Lemma 7.4; x_2 is regarded as a parameter.) By (4.21) and (4.23) the hypotheses of Lemma 7.4 are fulfilled, and $\tilde{F}_1$ is equivalent to $A_2\sqrt{\rho M}$ when $x_2 = 0$ by definition of A_2. (Here $M = M_0$.) (4.23) gives (see also (4.27))

$$|\partial(F(x_1, x_2, \xi_2) - F(x_1, 0, \xi_2))/\partial\xi_2| < C(\rho M)^{-1/2}B_2^{-1} = A_2\sqrt{\rho M}/M$$

so this is true for every x_2. It is more convenient for us to normalize F_1 so while keeping the notation

$$F_0(x_1, x_2) = F(x_1, x_2, 0) = \tilde{q}(x_1/M, x_2/B_2, 0)/\rho M$$

as in Lemma 7.4 we shall write

$$F_1(x_1, x_2) = (\partial F(x_1, x_2, 0)/\partial\xi_2)/A_2\sqrt{\rho M} = (\partial\tilde{q}(x_1/M, x_2/B_2, 0)/\partial\xi_2)(B_2/M)$$

which means that

$$L(x_1/M, x_2/B_2, \xi_2 B_2)/M = \rho F_0(x_1, x_2) + F_1(x_1, x_2)\xi_2 . \tag{6.17}$$

By (7.16) we have, after changing the sign of (x_2, ξ_2) if necessary,

$$F_1(x_1, x_2) \geqq -C/(A_2^2\rho M), \ |x_1| < 1, \ |x_2| < 1 , \tag{6.18}$$

and Lemma 7.4 also gives

$$|F_0(x_1, x_2)/F_1(x_1, x_2)| \leqq C \ \text{ if } \ F_1(x) > C/A_2^2\rho M, \ |x_1| < 1, \ |x_2| < 1 , \tag{6.19}$$

$$\begin{aligned} & F_0(x_1, x_2)/F_1(x_1, x_2) - F_0(y_1, x_2)/F_1(y_1, x_2) \leqq \\ & \leqq C(1/F_1(x_1, x_2) + 1/F_1(y_1, x_2))/A_2^2\rho M \ \text{ if } \ -1 < x_1 < y_1 < 1 , \end{aligned} \tag{6.20}$$

provided that the right hand side is small enough; then we have also

$$F_1(x_1, x_2)\partial(F_0(x_1, x_2)/F_1(x_1, x_2))/\partial x_1 \geqq -C(1 + |\partial F_1/\partial x_1/F_1|)/A_2^2\rho M . \tag{6.21}$$

Recall that $A_2 \geqq \varepsilon = \lambda^\kappa$, while $\sqrt{\rho M} \geqq R_2 = \lambda^{-\kappa_2}$ where $\kappa_2 > \kappa$, so $A_2\sqrt{\rho M} > \lambda^{(\kappa - \kappa_2)}$, the exponent being negative. For small λ it follows that the assumptions (5.41), (5.42) of Lemma 5.8 are fulfilled if $F = \rho F_0$ and $G = F_1$. By the remarks made after Lemma 7.4 this is still true if we take $G = F_1 + C/A_2^2 \rho M$, which makes G non-negative by (6.18). We have (5.37) by (4.28) and (5.38) follows from (4.23), (4.23)′. Since $M(x,\xi')/M$ is bounded from below in Ω_0 it follows as observed before the statement of Lemma 4.2 that (5.39) is fulfilled. If K is a compact set in R^2 contained in the unit square and containing the projection of the support of $\tilde{\Phi}_0(x_1/M, x_2/B_2, \cdots)$ there, we obtain from Lemma 5.8

$$\rho^{1/(k+1)}\|v\| \leqq C\,\|(D_1 + i(\rho F_0(x) + G(x)D_2))v\|, \quad v \in C_0^\infty(K)\,. \tag{6.22}$$

If we introduce the modified definition of G and change variables using (6.17) we obtain with $K' = \{(x_1/M, x_2/B_2);\ (x_1, x_2) \in K\}$

$$\rho^{1/(k+1)}\|Mv\| \leqq C(\|(D_1 + iL(x,D))v\| + \|A_2^{-1}D_2 v\|), \quad v \in C_0^\infty(K') \tag{6.23}$$

Here v is a function of two variables but we may of course let v depend also on $x_3, \cdots, x_n$ without any modification of the estimate.

In supp Φ_0 we have $|\xi_2/A_2| < R_2\varepsilon << M$. When we compose ξ_2/A_2 with the inverse of χ_a and apply the resulting operator to $\Phi_0 u$ we therefore obtain an error term of the same behavior as the error term r_j above, so it will play no role. Let us therefore examine how we can pass from (6.23) to an estimate where L is replaced by $L \circ \chi_a^{-1}$. We have essentially two kinds of canonical transformations to consider.

i) First assume that as on page 156

$$\chi_a(x',\xi') = (x_2, x_3 + f_3(x_2), \cdots, \xi_2 + g_2(x_2) + G(x',\xi'), \xi_3 + g_3(x_2), \cdots),$$

$$G(x',\xi') = \sum_3^n (g_j'(x_2)x_j - f_j'(x_2)\xi_j)\,.$$

The canonical transformation corresponding to the change of variables

$$y_2 = x_2,\ y_j = x_j + f_j(x_2),\ j > 2\,, \tag{6.24}$$

is defined by $\sum \eta_j \, dy_j = \sum \xi_j \, dx_j$, that is,

$$\xi_2 = \eta_2 + \sum_3^n \eta_j f'_j(x_2), \quad \xi_j = \eta_j \quad \text{for } j > 2 . \tag{6.25}$$

Thus

$$\chi_a(x', \xi') = (y_2, \cdots, y_n, \eta_2 + \partial H/\partial y_2, \cdots, \eta_n + \partial H/\partial y_n) ,$$

$$H(y) = \sum_3^n g_j(y_2)(y_j - f_j(y_2)) + \int g_2(y_2)\, dy_2 + \sum_3^n \int g_j(y_2)\, df_j(y_2) .$$

The change of variables (6.24) is measure preserving so it corresponds to a unitary transformation. To pass from (6.23) to the same estimate with the symbols on the right hand side composed with χ_a^{-1} we need only first replace v by ve^{iH}, which is a unitary transformation, and then change variables from y to x. Thus we obtain

$$\rho^{1/(k+1)} \| M_j \Phi_j(x, D') U_2 \| \leqq C(\| (P - ir_j(x, D')) \Phi_j(x, D') U_2 \| + \\ + \| A_2^{-1} \eta_2 \Phi_j(x, D') U_2 \|) . \tag{6.26}$$

Here η_2 denotes the composition of ξ_2 with χ_a^{-1}.

ii) Now we assume that $\chi_a = T_0 \circ \chi'_a$ where T_0 is the canonical transformation $(x', \xi') \to (\xi', -x')$ and χ'_a has the form considered in (i). Apart from trivial relabelling of coordinate pairs this hypothesis will be fulfilled if we cannot attain case i) in that way. We have now

$$L \circ \chi_a^{-1} = L' \circ T_0^{-1} \quad \text{where } L' = L \circ \chi_a'^{-1} .$$

Thus we have proved in part i) that

$$\rho^{1/(k+1)} \| M_0 v \| \leqq C(\| (D_1 + iL'(x, D)) v \| + \| A_2^{-1} \eta'_2 (x, D) v \|) \tag{6.27}$$

where η'_2 is the composition of η_2 with $\chi_a'^{-1}$. Actually it is preferable for reasons which will be clear in a moment to replace L in (6.23) by L_r, which is obtained by putting the coefficient of D_2 to the right of D_2. If we go back to (6.22) and note that $\partial G/\partial x_2$ has a fixed bound, we see immediately that this does not affect the validity of (6.22) or (6.23). Thus we may replace L' by L'_r in (6.27). Now we apply (6.27) to the Fourier transform of v in x. This changes $L'_r(x, D)$ to $L' \circ F_0^{-1}(x, D) = L \circ \chi_a^{-1}(x, D)$ so (6.27) gives

$$(6.28)\quad \rho^{1/(k+1)}\|M_0 v\| \leqq C(\|(D_1 + i(L \circ \chi_a^{-1})(x, D))v\| + \|A_2^{-1}\eta_2(x, D)v\|)\,.$$

In the preceding argument we neglected an important point, the condition supp $v \subset K$ in (6.23). This means that (6.28) is valid provided that $(Mx_1, B_2\xi_2)$ belongs to a fixed compact subset of $(-1, 1) \times (-1, 1)$ when (x, ξ) belongs to the support of the Fourier transform of $v(x_1, x')$ with respect to x'. This is not the case for $v = \Phi_0(x, D')U_2$. However, we can choose $h \in C_0^\infty(-1, 1)$ so that $h(B_2 x_2) = 1$ in supp $\tilde{\Phi}_0$. This means that $h(B_2\xi_2) = 1$ in supp Φ_0, so the symbol of $(h(B_2D_2) - 1)\Phi_0(x, D')$ divided by λ^N is of type R_2 for any N. If we apply (6.28) to $v = h(B_2D_2)\Phi_0(x, D')U_2$ which is legitimate, we therefore obtain for $j = 0$, thus for any j,

$$(6.26)'\quad \rho^{1/(k+1)}\|M_j\Phi_j(x, D')U_2\| \leqq C(\|(P - ir_j(x, D'))\Phi_j(x, D')U_2\| + \\ + \|A_2^{-1}\eta_2(x, D')\Phi_j(x, D')U_2\| + \lambda^N\|u\|)\,.$$

(6.26)′ completes our estimates for U_2. The discussion of U_1 is similar but without a number of the complications dealt with above, so we shall consider it very briefly. Instead of (6.14) we have

$$(6.14)'\qquad \|U_1\|^2 \leqq C\Big(\sum \|\phi_j(x, D')U_1\|^2 + \lambda^2\|u\|^2\Big)\,.$$

Let $P_j = D_1 + iq(x_1, x'_j, \xi'_j)$ where (x'_j, ξ'_j) are coordinates of the center of ω_j. Then (6.16) is replaced by

$$(6.16)'\quad \sum \|P_j\phi_j(x, D')U_1\|^2 \leqq C\Big(\|PU_1\|^2 + \sum \|m_j\phi_j(x, D')U_1\|^2 + \|u\|^2\Big)\,.$$

We can apply Lemma 5.3 to P_j and obtain

$$(6.26)'' \qquad \rho^{1/(k+1)}\|m_j\phi_j(x, D')U_1\| \leqq C\|P_j\phi_j(x, D')U_1\| .$$

If we recall (6.4), (6.10) and the preceding estimates, we have proved that

$$(\rho^{2/(k+1)} - C)\Big(\sum \|M_j\Phi_j(x, D')U_2\|^2 + \sum \|m_j\phi_j(x, D')U_1\|^2\Big) \leqq$$
$$\leqq C(\|Pu\|^2 + \|u\|^2) .$$

Now fix ρ so large that the parenthesis on the left hand side is positive. Then we obtain

$$\sum \|M_j\Phi_j(x, D')U_2\|^2 + \sum \|m_j\phi_j(x, D')U_1\|^2 \leqq C(\|Pu\|^2 + \|u\|^2) .$$

We can estimate M_j and m_j from below by $\lambda^{-2/(k+1)}$ according to (4.3), and by (6.14), (6.14)′ it follows that

$$(6.29) \qquad \lambda^{-2/(k+1)}\|U\| \leqq C(\|Pu\| + \|u\|) .$$

This is the asserted estimate (6.1) which completes the proof of Theorem 3.4.

In (6.23) we could have added a term $\|D_1 v\|$ on the left hand side and obtained a term $\|D_1 \Phi_j(x, D')U_2\|$ on the left hand side of (6.26)′. Similarly we can add a term $\|D_1\phi_j(x, D')U_1\|$ on the left hand side of (6.26)″, and this leads to the estimate

$$(6.30) \qquad \|D_1 U\| \leqq C(\|Pu\| + \|u\|) .$$

This contains the sufficiency of Theorem 3.4 in the sense that (6.30) combined with an argument based on the Campbell-Hausdorff formula which does not require any detailed decomposition of U gives (6.29). Shortcuts to (6.30) are sometimes available but not known in general.

7. *Calculus lemmas*

The following lemma is important in sections 2 and 4. It has been extracted from the proof of Lemma 6 in Egorov [3] and Lemma 2.2 in Egorov [2]; the special case Lemma 7.2 was already given by Louis Nirenberg in some informal talks in 1970.

LEMMA 7.1. *Let* F *be a real valued* C^∞ *function in*

$$\Omega = \{(t,y) \in \mathbf{R}^{1+N};\ |t| < 1 \ \textit{and}\ |y| < 1\}$$

such that with a fixed integer k

$$|F(t,0)| \leqq 1,\quad |\partial F(t,0)/\partial t| \leqq 1 \quad \textit{when}\ |t| < 1\ , \tag{7.1}$$

$$|\partial^{k+1} F(t,0)/\partial t^k \partial y| \leqq 1 \quad \textit{when}\ |t| < 1\ , \tag{7.2}$$

$$|\partial^2 F/\partial y^2| \leqq 1 \ \textit{and}\ |\partial^3 F/\partial t \partial y^2| \leqq 1 \ \textit{in}\ \Omega\ , \tag{7.3}$$

$$\partial F/\partial t \geqq 0 \quad \textit{when}\ F = 0\ . \tag{7.4}$$

Then there is a linear form $L(y)$ *and a polynomial* $G(t)$ *of degree* $< k$ *such that*

$$F(t,y) = G(t)L(y) + R(t,y) \tag{7.5}$$

$$|R(t,y)| + |\partial R(t,y)/\partial t| + |\partial R(t,y)/\partial y| + |\partial^2 R(t,y)/\partial y^2| \leqq C \quad \textit{in}\ \Omega\ , \tag{7.6}$$

where C *depends only on* N *and* k.

Proof. Application of Taylor's formula in y first and in t afterwards gives that

$$F(t,y) = \sum_1^N y_j F_j(t) + R(t,y)$$

where

$$F_j(t) = \sum_{i<k} t^i \partial^{i+1} F(0)/\partial y_j \partial t^i / i!,\ j = 1, \cdots, N\ ,$$

$$R(t,y) = F(t,0) + \int_0^1 (1-s)F''_{yy}(t,sy)\,yy\,ds + k\int_0^t (t-s)^{k-1}\partial^{k+1}F(s,0)/\partial y\partial s^k y\,\frac{ds}{k!}.$$

Since $\partial^2 R/\partial y^2 = \partial^2 F/\partial y^2$ we have

(7.7) $$|R| + |\partial R/\partial t| + |\partial R/\partial y| + |\partial^2 R/\partial y^2| \leqq C \text{ in } \Omega .$$

Let M be the maximum of the coefficients of $F_1, \cdots, F_N$. We may assume that M is large, for otherwise $\sum y_j F_j(t)$ could be included in R. Choose $t_0 \in (-1/2, 1/2)$ so that

$$\sum_1^N F_j(t_0)^2 > c\,M^2$$

for some c depending only on k. By a rotation of the y variables we can make $F_j(t_0) = 0$ for $j \neq 1$. The lemma follows if we prove that there is a fixed bound for F_j when $j \neq 1$ for $y_j F_j$ can be included in R then. In the proof we may as well assume that $N = 2$.

Consider the equations

(7.8) $$\sum_1^2 F_j(t)\,y_j = 0, \quad \sum_1^2 F'_j(t)\,y_j = -K$$

for some large constant K to be chosen later. We shall first examine if there is a solution in Ω and then if it can be modified to give a contradiction with (7.4). The solution of (7.8) is given by

(7.9) $$y_1 = KF_2(t)/H(t),\ y_2 = -KF_1(t)/H(t) \text{ where}$$
$$H(t) = F_1(t)F_2'(t) - F_1'(t)F_2(t) .$$

Choose δ with $0 < \delta < 1/2$ depending only on k so that

$$|F_1(t)| > c'\,M,\ \ |t - t_0| < \delta .$$

If $\tilde{F}_2$ is the maximum of the coefficients of F_2 we have

$$\tilde{F}_2 M \leqq C \max_{|t-t_0|<\delta} |H(t)| . \tag{7.10}$$

In fact, F_2 is the solution of the Cauchy problem

$$F'_2 - (F'_1/F_1) F_2 = (H/M)(M/F_1),\ F_2(t_0) = 0 .$$

Since we have uniform bounds for the derivatives of F'_1/F_1 and M/F_1 at t_0, the derivatives of F_2 can be estimated by those of H/M. Choose $t \in (t_0 - \delta, t_0 + \delta)$ so that

$$\tilde{F}_2 M \leqq C |H(t)| . \tag{7.10$'$}$$

Then (7.9) implies that with another C

$$|y_1| < CK/M,\ |y_2| < CK/\tilde{F}_2 . \tag{7.11}$$

If $\tilde{F}_2 > 2CK$, hence $M > 2CK$, then $|y_j| < 1/2$ for $j = 1, 2$. Now we obtain from (7.7) and (7.8) if $\varepsilon < 1/4$

$$|F(t, y_1 \pm \varepsilon, y_2) \mp F_1(t)\varepsilon| < C_1,\ |F'_t(t, y_1 \pm \varepsilon, y_2) \mp F'_1(t)\varepsilon + K| < C_1 .$$

If M is so large that $|F_1(t)|\varepsilon = C_1$ implies $\varepsilon < 1/4$, we conclude that there is some $Y_1 \in (y_1 - \varepsilon, y_1 + \varepsilon)$ such that $(t, Y_1, y_2) \in \Omega$ and $F(t, Y_1, y_2) = 0$. Since $|F_1(t)|\varepsilon = C_1$ implies $|F'_1(t)\varepsilon| \leqq C_2$ we obtain $F'_t(t, Y_1, y_2) < 0$ if $K = C_1 + C_2$. This contradicts (7.4) so we conclude that

$$\tilde{F}_2 \leqq 2C(C_1 + C_2) ,$$

which proves the lemma.

REMARK. The linear form L in (7.5) may be replaced by any form L_1 with $\|L - L_1\| < C/\tilde{G}$, where $\tilde{G}$ is the smallest bound for the coefficients of G, for $G(t)(L(y) - L_1(y))$ may then be included in R. By (7.5) we have if (7.1) is supplemented by a bound for $\partial^k F(t, 0)/\partial t^k$

$$|<d_y F^{(s)}(0,0), y> - G^{(s)}(0)L(y)| < C, \ |y| < 1 .$$

If we choose s so that $0 \leqq s < k$ and $|d_y F^{(s)}(0,0)|$ is maximal, and if the maximum is large, we obtain $\tilde{G} \leqq C'_k |G^{(s)}(0)|$. Thus we may take

$$L_1 = <d_y F^{(s)}(0,0), y>/G^{(s)}(0) .$$

By a rotation of the coordinates we can make L_1 proportional to y_1. Then it follows from Taylor's formula as in the proof of Lemma 7.1 that the difference between $L_1(y)\,G(t)$ and the Taylor expansion of $y_1 \partial F(t,0)/\partial y_1$ of order $k-1$ in t has a uniform bound, so (7.6) is valid with R replaced by $F(t,y) - y_1 \partial F(t,0)/\partial y_1$.

Lemma 7.1 gives immediately information on the Taylor expansion of functions satisfying (7.4):

LEMMA 7.2. *Let* F *be a* C^∞ *function of* $(t,y) \in \mathbf{R}^{1+N}$ *in a neighborhood of* $(0,0)$, *assume that* F *satisfies* (7.4) *and that* $F(t,0) = \mathcal{O}(t^\mu)$. *Then there is a linear form* L *in* $\mathbf{R}^N$ *and a polynomial* $G(t)$ *of degree* $< \mu/2$ *such that*

$$F(t,y) = G(t)\,L(y) + \mathcal{O}(|t|^\mu + |y|^2) .$$

Proof. Set

$$F_\varepsilon(t,y) = \varepsilon^{-\mu} F(\varepsilon t, \varepsilon^{\mu/2} y) .$$

Then (7.1)-(7.3) are satisfied uniformly in ε when $\varepsilon \to 0$ if $k \geqq \mu/2$. Hence there is a linear form L_ε with norm 1 say such that

$$\varepsilon^{i-\mu/2} |<\partial^{i+1} F(0)/\partial t^i \partial y, y>| < C \ \text{ if } \ L_\varepsilon(y) = 0 .$$

If L is a limit of L_ε it follows that $<\partial^{i+1}F(0)/\partial t^i \partial y, y> = 0$ if $L(y) = 0$ and $i < \mu/2$, which proves Lemma 7.2.

The following observation is useful in section 2:

LEMMA 7.3. *If* $F_j \in C^1(\Omega)$ *satisfies* (7.4) *and* $F_j \to F$ *in* $C^1(\Omega)$ *as* $j \to \infty$, *then* F *satisfies* (7.4).

Proof. Assume that $\partial F/\partial t < 0$ at a zero $(t_0, y_0) \in \Omega$ of F. By the implicit function theorem the equation $F_j(t,y) = 0$ has a zero close to (t_0, y_0) with $\partial F_j(t,y)/\partial t < 0$ if j is large, which is a contradiction proving the lemma.

We shall now examine further the properties of $F(t,0)$ and $G(t)$ in Lemma 7.1 when (7.4) is strengthened to

(7.12) If $F(t,y) > 0 > F(s,y)$ and $(t,y), (s,y) \in \Omega$, then $s < t$.

There is no point now in having many y variables so we assume $N = 1$. We write

$$F_0(t) = F(t,0), \quad F_1(t) = \partial F(t,0)/\partial y .$$

Then with a slightly changed definition of R

$$F(t,y) = F_0(t) + F_1(t)\,y + R(t,y) \tag{7.13}$$

where by Taylor's formula and (7.3)

$$|R(t,y)| + |\partial R(t,y)/\partial t| \leqq y^2, \quad |\partial R(t,y)/\partial y| \leqq |y| . \tag{7.14}$$

It is very easy to see that either $F_1 \geqq -2$ in $(-1,1)$ or else $F_1 \leqq 2$ in $(-1,1)$. In fact, assume that $F_1(t) > 2$ for some t. Then

$$F(t,y) = F_0(t) + F_1(t)\,y + R(t,y)$$

is positive if y is close to 1 and negative if y is close to -1, so (7.12) shows that $F(s,y) \geqq 0$ if $s \geqq t$ and y is close to 1 while $F(s,y) \leqq 0$ if $s \leqq t$ and y is close to -1. Thus $F_1(s) + 2 \geqq 0$ if $s \geqq t$ and $-F_1(s) - 2 \leqq 0$ if $s \leqq t$, which proves that $F_1 \geqq -2$. Changing the sign of y if necessary we may therefore always assume $F_1 \geqq -2$.

Since $|F_1^{(k)}| \leqq 1$ we know that F_1 differs from a polynomial of degree $k-1$ by a function which is bounded in C^k. We assume that

$$\tilde{F}_1 = \max_{j<k} |F_1^{(j)}(0)|$$

is so large that the polynomial part dominates. We can choose $t \in (-1,-1/2)$ so that

$$\tilde{F}_1 < C_1 F_1(t)$$

where C_1 only depends on k. Then

$$F(t, y) = F_0(t) + F_1(t) y + R(t, y)$$

has the sign of y if $|y| \tilde{F}_1/C_1 > 2$. We can find $t \in (1/2, 1)$ with the same property and conclude from (7.12) that $F(s, y)$ has the sign of y if $|s| < 1/2$ and $|y| > 2C_1/\tilde{F}_1$. Hence

$$|F_0(s)| \leqq 2C_1 F_1(s)/\tilde{F}_1 + 4(C_1/\tilde{F}_1)^2, \quad |s| < 1/2, \tag{7.15}$$

which in particular implies that

$$F_1(s) \geqq -2C_1/\tilde{F}_1, \quad |s| < 1/2. \tag{7.16}$$

Note that (7.15) shows that

$$|F_0(s)| \leqq 4C_1 F_1(s)/\tilde{F}_1 \quad \text{if} \quad F_1(s) > 2C_1/\tilde{F}_1, \quad |s| < 1/2. \tag{7.15$'$}$$

We shall now prove that F_0/F_1 is essentially increasing. To do so we take s with $|s| < 1/2$ so that

$$K = |F_1(s)| \tilde{F}_1$$

is large. Then we can use (7.15)$'$ so if $y = -F_0(s)/F_1(s)$ we have $|y|\tilde{F}_1 \leqq 4C_1$. If $|z| < |y|$ we obtain

$$F(s, y+z) = F_1(s) z + R(s, y+z) = F_1(s)(z + \mathcal{O}(|y|/K)).$$

The parenthesis has opposite signs for $z = \pm C|y|/K$ if C is large enough. Choose z so that $F(s, y+z) < 0$. If $t < s$ it follows that $F(t, y+z) \leqq 0$, so

$$F_0(t) + F_1(t)(-F_0(s)/F_1(s) + z) \leqq (y+z)^2 \leqq C'/\tilde{F}_1^2 ,$$

and $|z| < C/|F_1(s)|\tilde{F}_1^2$. Hence

$$F_0(t)/F_1(t) - F_0(s)/F_1(s) \leqq C_2(1/F_1(s) + 1/F_1(t))/\tilde{F}_1^2 \tag{7.17}$$
$$\text{if } -1/2 < t < s < 1/2 \text{ and } \tilde{F}_1 \min(F_1(s), F_1(t)) \text{ is large.}$$

In the preceding discussion we could have chosen z with $|z| < C|y|/K$ so that $F(s, y+z) = 0$. In view of (7.4) we obtain

$$0 \leqq \partial F(s, y+z)/\partial s \leqq F'_0(s) + F'_1(s)(y+z) + C/\tilde{F}_1^2$$

which gives

$$F_1(s) \frac{d}{ds}(F_0(s)/F_1(s)) > -C_3(1 + |F'_1(s)/F_1(s)|)/\tilde{F}_1^2 \text{ if} \tag{7.18}$$
$$|s| < 1/2 \text{ and } F_1(s)\tilde{F}_1 \text{ is large enough.}$$

Summing up, we have proved

LEMMA 7.4. *Let* F *satisfy the hypotheses of Lemma* 7.1 *with* $N = 1$ *and also* (7.12). *Set* $F_j(t) = \partial^j F(t, 0)/\partial y^j$, $j = 0, 1$. *If* F_1 *is sufficiently large at some point in* $(-1, 1)$ *it follows that* (7.15) - (7.18) *are valid.*

In the preceding inequalities one may replace F_1 by $G = F_1 + 2C_1/\tilde{F}_1$ which is non-negative by (7.16). In fact, $\tilde{G}/\tilde{F}_1$ is close to 1 if $\tilde{F}_1$ is large, and $G\tilde{G}$ is large if and only if $F_1\tilde{F}_1$ is large. It is clear that replacing F_1 by G in the right hand side of (7.15) only requires a change of constants. This is also true of (7.17) since

$$|F_0(t)|\,|1/F_1(t) - 1/G(t)| < 2C_1|F_0(t)|/\tilde{F}_1|F_1(t)G(t)| < C/G(t)\tilde{F}_1^2 .$$

Replacing F_1 by G in (7.18) changes the left hand side by

$$F_0 G'(1/F_1 - 1/G)$$

which we can estimate by $|G'/G|\,\tilde{F}_1^{-2}$. Hence (7.18) also remains valid.

8. *Concluding remarks*

We shall now indicate briefly the relations between this paper and Egorov [2, 3]. Section 2 here differs little in contents from Egorov [2, Chap. II, §1]. However, our use of weights makes the discussion easier, and it fits perfectly with the proof of necessary conditions in section 3, corresponding to Egorov [2, Chap. III, §1]. Also (2.17) seems to be new, for Egorov remarks [2, p. 83] on (2.13) that "we do not use this fact [but it] may turn out to be useful"; on the other hand (2.17) identifies (2.13) with the basic condition on forbidden sign changes. Section 4 is essentially an exposition of Egorov [3, §6-7] until we come to the definition of the neighborhood Ω_0 in case II. We have given it a standard width R_2 in the ξ_2 direction while Egorov defines this width to be $1/A_2$ which has the advantage that $M(x, \xi')$ becomes essentially constant in Ω_0, just as in case I. However, the fact that Ω_0 is so thin, added to the difficulty of the canonical transformations χ_a makes his partition of unity Φ_j inaccessible to any standard calculus of pseudo-differential operators. Egorov [3, §13-14] gives a discussion of this problem which is hard to follow, and we have seen no way of justifying it except by means of a double partition of unity such as used here.

The partition of unity we introduce in section 4 belongs to a well established class of pseudo-differential operators. However, the fact that we make the neighborhoods Ω_j so large requires us to choose their centers carefully; this is the reason for the notion of admissible neighborhood in section 4. In addition we must give in section 5 a much more complete study of the localized estimates. The methods we use there depend on Egorov [3, §10]. He uses them essentially to prove Lemma 5.6. We have also used them to prove the estimate (Lemma 5.3) required in case I (see also Treves [10] and for a somewhat incomplete proof Egorov [2, p. 90-95]). By combining these results we have proved the much stronger Lemma 5.8 which is the tool allowing us to use the large neighborhoods Ω_j. The proof of Lemma 5.8 parallels that used in section 6 to finish the proof of sufficiency in Egorov's theorem. This duplication may not only be a disadvantage for it means that one encounters the ideas of the proof first in a much more elementary case.

REFERENCES

[1] Egorov, Yu. V., Subelliptic pseudo-differential operators. Dokl. Akad. Nauk SSSR 188 (1969), 20-22. Also Soviet Math. Dokl. 10 (1969), 1056-1059.

[2] ———, Subelliptic operators. Usp. Mat. Nauk 30:2 (1975), 57-114; Russian Math. Surveys 30:2 (1975), 59-118.

[3] ———, Subelliptic operators. Usp. Mat. Nauk 30:3 (1975), 57-104; Russian Math. Surveys 30:3 (1975), 55-105.

[4] Hörmander, L., Pseudo-differential operators and non-elliptic boundary problems. Ann. Math. 83 (1966), 129-209.

[5] ———, Hypoelliptic second order differential equations. Acta Math. 119 (1967), 147-171.

[6] ———, Propagation of singularities and semi-global existence theorems for (pseudo-)differential operators of principal type. Ann. of Math. 108 (1978).

[7] ———, Spectral analysis of singularities. Seminar on Singularities of Solutions of Linear Partial Differential Equations, Princeton University Press 1979, 3-49.

[8] Nagano, T., Linear differential systems with singularities and an application to transitive Lie algebras. J. Math. Soc. Japan 18 (1966), 398-404.

[9] Nirenberg, L., and Treves, F., On local solvability of linear partial differential equations. Part I: Necessary conditions. Comm. Pure Appl. Math. 23 (1970), 1-38.

[10] Treves, F., A new method of proof of the subelliptic estimates. Comm. Pure Appl. Math. 24 (1971), 71-115.

LACUNAS AND TRANSMISSIONS

Louis Boutet de Monvel

The purpose of this lecture is to describe the method of L. Gårding [1] for the study of lacunas of solutions of hyperbolic equations. For this a first step is the study of symmetries (transmissions) of Fourier integral distributions; such symmetries have been studied systematically by A. Hirschowitz and A. Piriou [2], and it is their point of view I present here. Transmissions also appear in pseudo-differential elliptic boundary value problems.

1. *Sharp fronts*

Let X be an open set in $\mathbf{R}^n$ (or a C^∞ manifold). A lacuna of a distribution f on X is an open set in which f vanishes. Pseudo-differential methods, in which one is led to neglect C^∞ functions, are not quite adapted to the study of lacunas in this sense, but may give good results about the sharp fronts which are defined as follows:

DEFINITION 1. *Let* f *be a distribution on* X, $U \subset X$ *an open set, and* x_0 *a point of the boundary* ∂U *of* U *in* X. *We will say that* f *is sharp at* x_0, *from* U, *if there exists* $g \in C^\infty(X)$ *such that* $f - g$ *vanishes in* $U \cap V$ *for some neighborhood* V *of* x_0.

If f is not sharp, we say that it is diffuse. For the study of sharp

Seminar on Singularities of Solutions
0-691-08221-9/78/0209 \$00.50/1 (cloth)
0-691-08213-8/78/0209 \$00.50/1 (paperback)

fronts of distributions defined by an oscillatory integral, the basic remark is the following.

PROPOSITION 1. *Let* $a(\xi) \sim \sum_{k=0}^{\infty} a_{s-k}(\xi)$ *be a regular*(*) *symbol of degree* $s\,(s \in \mathbb{C})$ *in one variable* ξ, *and let* $f(x) = \frac{1}{2\pi} \int_{-\infty}^{\infty} e^{ix\cdot\xi} a(\xi)\, d\xi$ *be the inverse Fourier transform. Then* f *is sharp at* 0 *from the right if for each* k *we have*

$$a_{s-k}(-\xi) = e^{si\pi}(-1)^k a_{s-k}(\xi) \quad \textit{for} \ \xi > 0\,. \tag{1.1}$$

The condition means that the homogeneous function $a_{s-k}(\xi)$ has a holomorphic extension in the upper complex half plane $\operatorname{Im} \xi > 0$, so that it is the Fourier transform of a distribution with support in the negative half-line, and the proposition follows immediately.

Obviously this result allows parameters, so it extends to distributions $f \in I_\Lambda^t$ where Λ is the conormal bundle of a hypersurface $Y \subset X$. (We recall that if Λ is a Lagrangian cone in T^*X, I_Λ^t denotes the set of Fourier integral distributions of degree t associated with Λ, with regular symbol. Locally these distributions have an integral representation:

$$f = \int e^{i\phi(x,\theta)}\, a(x,\theta)\, d\theta$$

where ϕ is a generating function for Λ, and a a regular symbol of degree $s = t + \frac{n_x}{4} - \frac{n_\theta}{2}$ (where n_x (resp. n_θ) is the number of x (resp. θ) variables):

$$a \sim \sum a_{s-k}(x,\theta)$$

with a_{s-k} homogeneous of degree s–k with respect to θ. The total

(*) or "classical," i.e. a_{s-k} is homogeneous of degree s–k for each *integer* k.

degree s may be any complex number, but for the transmission property below it is important that the degrees go down by integral steps.)

It is also of interest to recall what happens when f is not sharp in Proposition 1. The picture is slightly different according as the degree s of the symbol a is or is not an integer. If s is not an integer, the inverse Fourier transform f is of the form

$$f = \mathrm{pf}(|x|^{-s-1}(g_+(x)+g_-(x))) + C^\infty \tag{1.2}$$

where $g_\pm(x)$ is C^∞ for $\pm x \geq 0$ (up to the origin), and vanishes on the other side (pf denotes a Hadamard finite part); f is sharp from the right if $g_+ = 0$ (or if g_+ vanishes of infinite order for $x = 0$).

If s is an integer and f is sharp from one side, it is also sharp from the other side and has the form

$$f = g_+(x) + g_-(x) + \sum_{j \leq s} c_j \delta^{(j)} \quad (c_j = \text{constants}) \tag{1.3}$$

where g_+, g_- are as above (their Taylor series at 0 must agree to the order $-s-1$ if $s < 0$).

If a has the opposite symmetry: $a_{s-k}(-\xi) = -(-1)^{s-k} a_{s-k}(\xi)$, f has the form

$$f = a(x)\,\mathrm{pf}\, x^{-s-1} + b(x) \log|x| + C^\infty \tag{1.4}$$

with $a, b \in C^\infty$ (b must vanish of order $-s-1$ if $s < 0$). So if f is not sharp, it is diffuse from both sides.

Figure 1 below shows the singularity in the cases which will be most useful later: in case I±, s is an integer and $a_{s-k}(-\xi) = \pm(-1)^{s-k} a_{s-k}(\xi)$. In case II±, s is a half-integer ($s \in \mathbf{Z} + \frac{1}{2}$) and $a_{s-k}(-\xi) = \pm e^{(s-k)i\pi} a_{s-k}(\xi)$ $(\xi > 0)$.

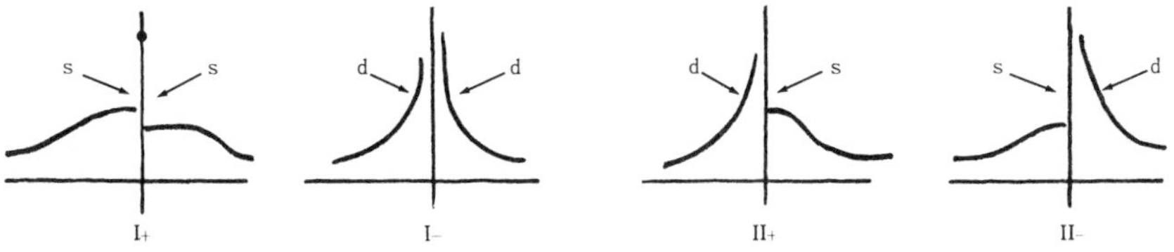

Fig. 1

2. *Transmissions*

Proposition 1 relates the sharpness to a symmetry condition on a symbol. Such symmetries are studied systematically in [2]. We will denote by

$$p \mapsto \tilde{p} \sim \sum (-1)^{m-k} p_{m-k}(x, -D) \tag{2.1}$$

the fundamental involution of the algebra of pseudo-differential operators (here the degree m of $p = p(x, D)$ is an integer; $\tilde{p}$ is only defined up to an operator of degree $-\infty$). We will say that a Lagrangian cone $\Lambda \subset T^*X \backslash 0$ is symmetric if $\Lambda = -\Lambda$ (i.e. $(x, \xi) \in \Lambda \Longleftrightarrow (x, -\xi) \in \Lambda$). Let Λ be symmetric: a transmission is an involution τ of I^t_Λ (mod. $I^{-\infty}_\Lambda$) such that $\tau(pf) = \tilde{p}(\tau f)$ (mod. C^∞) for each $f \in I^t_\Lambda$. Locally Λ consists of two pieces Λ_+ and $\Lambda_- = -\Lambda_+$, defined by opposite phase functions $\phi, -\phi$, and any $f \in I^t_\Lambda$ has a representation

$$f = \int (e^{i\phi(x,\theta)} a^+(x, \theta) + e^{-i\phi(x,\theta)} a^-(x, \theta))\, d\theta \tag{2.2}$$

PROPOSITION 2. *Any transmission* τ *on* I^t_Λ *is locally of the form*

$$f = \int (e^{i\phi} a^+ + e^{-i\phi} a^-)\, d\theta \mapsto \tau f = \int \left(e^{i\phi} \frac{\tilde{a}_-}{\lambda} + e^{-i\phi} \lambda \tilde{a}_+\right) d\theta$$

where λ *is a complex number,* $\lambda \neq 0$.

We have set $\tilde{a}(x, \theta) \sim \Sigma(-1)^k a_{s-k}(x, \theta)$, where s is the degree of a $\left(s = t + \frac{n_x}{4} - \frac{n_\theta}{2}\right)$. The proposition follows easily from the symbolic calculus of pseudo-differential operators and Fourier integral distributions.

If $f \in I^t_\Lambda$, we will say that it is symmetric, or that it has the transmission property, if $f = \tau f$ (mod. C^∞) for some transmission τ (such a τ is unique if $f \notin C^\infty$). Similarly we will say that a Fourier integral operator is symmetric (or has the transmission property) if its Schwartz-kernel is symmetric; for example a differential operator P (polynomial) is always symmetric $(P = \tilde{P})$, and so is its parametrix if P is elliptic. If a Fourier integral operator A and a distribution f are both symmetric, then so is Af. It is also immediate to check that the transmission property propagates: if P is symmetric $(P = \tilde{P})$ and has real simple characteristics, if $f \in I^s_\Lambda$ and $Pf \in C^\infty$, and f is symmetric at some characteristic point (x_0, ξ_0), then this symmetry propagates along the bicharacteristic strip through (x_0, ξ_0).

The complex number λ which enters in Proposition 2 is uniquely defined (and locally constant). It depends on the choice of ϕ; precisely if ϕ' is another phase function defining Λ, we have (locally) another integral representation of f;

$$f = \int (e^{i\phi'} a'_+ + e^{-i\phi'} a'_-)\, d\theta'$$

where the new leading terms are related to the old ones by

$$\sigma(a'_+) = J \exp\left(\frac{\mu i \pi}{4}\right) \sigma(a_+), \quad \sigma(a'_-) = J \exp\left(-\frac{\mu i \pi}{4}\right) \sigma(a_-)$$

where J is the square root of the absolute value of a Hessian determinant (the same for a_+ and a_-), and μ the signature of a Hessian matrix, describing the Maslov cocycle. We then get

$$\lambda' = i^{-\mu} \lambda . \tag{2.3}$$

Since $\mu = n_\theta - n_{\theta'}$ mod. 2, λ is completely determined up to a sign by τ and by the number of θ-variables. At this point appears the square $(-1)^\mu$ of the Maslov cocycle i^μ: this must be trivial if there exists a global transmission on I^s_Λ.

3. *Boundary value problems*

Let $\Omega \subset X$ have a smooth boundary $\partial\Omega$, and let $f \in I^s_\Lambda$ where $\Lambda = N(\partial\Omega)$ is the conormal bundle of $\partial\Omega$: f has an integral representation

(3.1) $$f = \int_{-\infty}^{+\infty} e^{itu(x)} a(x,t)\,dt$$

where u is a defining function for Ω ($u>0$ in Ω, $u=0$ on $\partial\Omega$, and $du \neq 0$ on $\partial\Omega$) and a is a regular symbol (of degree $m = s + \frac{n_x}{4} - \frac{1}{2}$). Then as in $n^0 1$ f is sharp from the interior along $\partial\Omega$ if

(3.2) $$a(x,-t) \sim e^{mi\pi} \sum (-1)^k a_{m-k}(x,t) \quad \text{for} \quad t>0$$

f is sharp from the exterior if we have the same relation, with $e^{-mi\pi}$ instead of $e^{mi\pi}$.

Let f be sharp from the exterior. Then inside of Ω we have $f = u^{-m-1} g$ (mod. C^∞), with $g \in C^\infty(\bar\Omega)$ ($m = s + \frac{n_x}{4} - \frac{1}{2}$). If P is a pseudo-differential operator of degree μ, Pf is sharp from the interior along $\partial\Omega$ if we have

(3.3) $$P_{\mu-k}(x,-\xi) = e^{(2m+\mu)i\pi}(-1)^k P_{\mu-k}(x,\xi)$$

in a conic neighborhood of the positive conormal bundle $N^+(\partial\Omega)$ (consisting of positive multiples of du, with u as above) (in fact it is enough that the Taylor series agree along $N^+(\partial\Omega)$).

This remark makes it possible to define elliptic (correctly posed) boundary value problems for operators such as $\sqrt{-\Delta}$: for instance one will look for a solution f of $\sqrt{-\Delta}\, f = g \in C^\infty(\bar\Omega)$, of the form $f = u^{k+1/2} f_1$ (k an integer) with $f_1 \in C^\infty(\bar\Omega)$. The 'classical' boundary conditions are replaced by pseudo-differential conditions on the Cauchy data of f_1, and in this example it is possible to choose them so that the problem behaves like an elliptic boundary value problem. Such problems were studied by L. Hörmander [3], and it is in this context that the transmission property appeared first. In [4] I described systematically such problems, but only in the case where the data g and the solution f are required to be sharp from both sides.

4. *Symmetry of the elementary solution of a hyperbolic equation*

The symmetries that occur most frequently in PDE theory, and in particular in connection with hyperbolic equations, are of a somewhat more restrictive type: distributions having this symmetry are locally of the form

$$f = \int_{R^N} e^{i\phi(x,\theta)} a(x,\theta)\,d\theta \tag{4.1}$$

where ϕ is symmetric $(\phi(x,-\theta) = -\phi(x,\theta))$ and a has the symmetry of a polynomial, i.e. its degree m is an integer, and

$$a \sim \sum a_{m-k}(x,\theta) \quad \text{with} \quad a_{m-k}(x,-\theta) = (-1)^{m-k} a_{m-k}(x,\theta) \tag{4.2}$$

(or more generally f is a superposition:

$$f = \sum_{j=1}^{M} \int e^{i\phi_j(x,\theta)} a_j(x,\theta)\,d\theta \tag{4.1 bis}$$

where the ϕ_j go by symmetric pairs, e.g. $\phi_{M+1-j}(x,\theta) = -\phi_j(x,\theta)$, and for the corresponding symbols $a_j \sim \Sigma\, a_{j,m-k}$, we have

$$a_{M+1-j,m-k}(x,-\theta) = (-1)^{m-k} a_{j,m-k}(x,\theta)\ .) \tag{4.2 bis}$$

If $P(x,t,D_x,D_t)$ is a differential operator of degree m, strictly hyperbolic with respect to t, and if E is the elementary solution at x_0:

$$P(E) = 0, \left(\frac{\partial}{\partial t}\right)^k E(x,0) = \begin{cases} 0 & \text{if } k < m-1 \\ \delta(x-x_0) & \text{if } k = m-1\ . \end{cases} \tag{4.3}$$

E always has this type of symmetry, with $n_\theta = n_x$, the degree of the symbols in (4.1) bis being $-\deg P + 1$. (This is at least true near $t = 0$, and globally if the bicharacteristic strips of P are complete, since the symmetry propagates.) For instance if $P = P(D_x, D_t)$ has constant coefficients on R^{n+1}, then E has a representation

$$E = \sum_1^m \int_{R^n} e^{i(x\cdot\xi + t\psi_j(\xi))} a_j(\xi)\, d\xi \tag{4.4}$$

where the ψ_j are the roots of the characteristic equation $P_m(\xi, \psi) = 0$ arranged in increasing order, and the a_j are symbols of degree $-m+1$, with a symmetry as above.

We now go back to the study of (4.1). Using the rule of $n^0 2$, we see that for any other representation of f as a Fourier integral (with symmetric phase), the complex number λ occurring in Proposition 2 is either ± 1, the degree of the symbol being an integer, or $\pm i$, the degree of the symbol being a half integer.

Let us call regular point of the Lagrangian cone Λ a point at which the projection $(x, \xi) \mapsto x$ is of maximal rank $n_x - 1$. Near such a point the projection of Λ is a hypersurface $Y \subset X$, and f has one of the singularities described in section 1, Figure 1. In particular for the elementary solution of a hyperbolic equation, this gives the well-known result that the singularity is always one of those described in case I or case II, according as the number n_x of variables of space type is odd or even.

Let us examine in particular the case where f is defined on R^n by

$$f = \int_{R^n} e^{i(x\cdot\xi + \psi(\xi))} a(\xi)\, d\xi .$$

with $\psi(-\xi) = -\psi(\xi)$, a having the symmetry of a polynomial. The critical Lagrangian manifold Λ is then defined by the equation $x = -\frac{\partial\psi}{\partial\xi}$, and the condition that the projection $\Lambda \to R^n$ is of maximal rank $n-1$ at a point (x_0, ξ_0) is that the Hessian matrix $\frac{\partial^2\psi}{\partial\xi^2}$ be of (maximal) rank $n-1$.

Then near x_0 the projection $Y = \Pi(\Lambda)$ is a hypersurface, defined by an equation of the form $<x-x_0, \xi_0> = g(x)$ where g vanishes of order 2 at x_0, and the Hessian of its restriction to the hyperplane $<x-x_0, \xi_0> = 0$ has the same signature as that of ψ (e.g. if $\psi = \sum_2^n \pm \xi_1^2/2\xi_1$, Y is defined by $x_1 = \frac{1}{2} \sum_2^n \pm x_j^2$). The rule describing the singularity of f is quite simple: we are in case I or case II according as $n_x - 1$ is even or odd; if Y is strictly convex, f is sharp from the outside of the convexity, and from there one alternates from case I+ to I– (resp. II+ to II–) according to the number of negative eigenvalues of the Hessian of g above. For example if E is the elementary solution of a 3rd order hyperbolic equation with 2 space variables $(n_x = 2)$ we get the following picture for the singularity, for fixed $t \neq 0$:

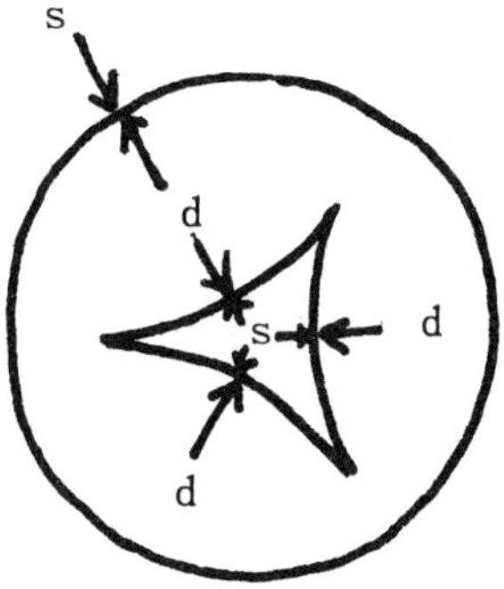

Fig. 2

The same analysis shows also quite well how the singular picture may change through a cusp point: possible cases are as follows.

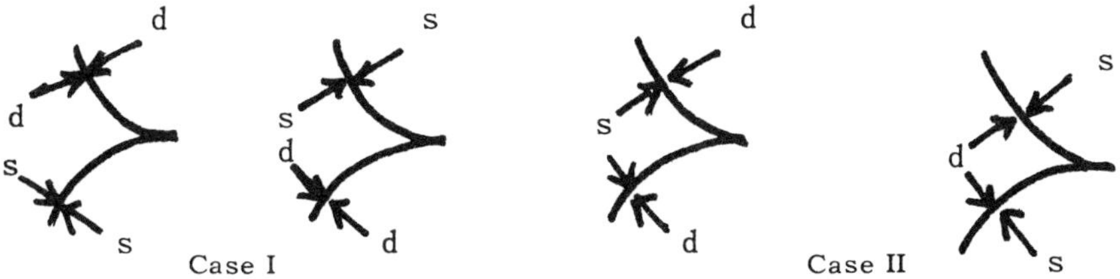

Fig. 3

(where s indicates that the singularity is sharp, d that it is diffuse). Case I gives an example of what Maslov calls a metamorphosis of the singularities. Similar examples are discussed in detail in [2].

5. *Further developments*

To go further it is necessary to develop methods to determine if a distribution f is sharp at the projections of points of Λ which are not regular, such as the cusp point of Figure 3. This is what L. Gårding does in [1], and I will not repeat his argument here. The method of Gårding consists in integrating first in the radial direction, then to regroup terms according to symmetries (there are 4 possibilities, as in Figure 1), and to push the resulting integration cycle into the complex domain (the new integration cycle is called a Petrovsky cycle). It may then happen that the new complex integral still converges over the singularity, thus exhibiting a C^∞ extension. This method, combined with the convexity rule above, gives easy rules to determine the sharpness when there are at most two θ-variables, and in particular at a cusp point, or at the origin of a swallow's tail.

REFERENCES

[1] Gårding, L., Sharp fronts of paired oscillatory integrals. Publ. R.I.M.S., Kyoto, 12(1977), 53-68.

[2] Hirschowitz, A., Piriou, A., Transmissions pour les distributions integrales de Fourier, to be published.

[3] Hörmander, L., Seminar, I.A.S. 1965

[4] Boutet de Monvel, L., Boundary problems for pseudo-differential operators. Acta Math. 126(1971), 11-51.

SOME CLASSICAL THEOREMS IN SPECTRAL THEORY REVISITED

Victor Guillemin[1)]

0. *Introduction*

Let D be a smooth strictly convex region in R^2 containing the origin. Let $\lambda D = \{\lambda x, x \epsilon D\}$. A classical theorem of Van der Corput says that the number of lattice points in λD is equal to λ^2 volume $D + O(\lambda^{2/3})$. This theorem is the simplest and most transparent of the "Weyl-type" theorems to which I am consecrating these lectures. In the first lecture, I will discuss this theorem and generalizations of it by Randol and Colin de Verdière. For me this result is intriguing because it already involves, in an elementary way, difficulties with "periodic bicharacteristics" of the kind one finds cropping up in a lot of recent work in spectral theory.

A classical theorem of Hermann Weyl [38] and Courant [9] says that if D is a bounded domain in R^2 and $N(\lambda)$ the number of eigenvalues $\leq \lambda^2$ of the Dirichlet problem: $\Delta u + \lambda u = 0$ with $u = 0$ on ∂D; then $N(\lambda) = (\text{volume } D/4\pi)\lambda^2 + O(\lambda \log \lambda)$. Recently efforts have been made to generalize this to manifolds with boundary and to improve the error term. In the second lecture I will describe these efforts. Most of the material for this lecture comes from an unpublished paper of Bob Seeley: "A sharp asymptotic remainder estimate for the spectrum of the Laplacian in a domain in R^3." I'm grateful to him for letting me reproduce them here.

---

1) Supported in part by NSF grant MCS77-18723.

Seminar on Singularities of Solutions
0-691-08221-9/78/0219 $02.05/1 (cloth)
0-691-08213-8/78/0219 $02.05/1 (paperback)

The third lecture in this series is devoted to a theorem of Szegö. To describe it I'll need a couple of preliminary definitions: Let V be a finite dimensional Hilbert space and $A: V \to V$ a self-adjoint linear operator. Then, for every function, f, on the real line, we can form $f(A)$ by means of the spectral theorem. The mapping which takes f into trace $f(A)$ can be thought of as a continuous functional on the space $C(\mathbf{R})$; so by the Riesz representation theorem it defines a measure, μ_A, on the real line which I'll henceforth call the *spectral measure* of A.(*) Of course μ_A is just $\Sigma\, \delta(\lambda - \lambda_i)$, $\lambda_i \in \operatorname{spec} A$. If V is infinite dimensional, we can still attempt to define a measure by the formula

$$\mu_A(f) = \operatorname{trace} f(A) \;;$$

however, in general the right hand side won't be defined. Nevertheless, in some instances one can get around this difficulty by defining the right hand side as a limit of finite dimensional traces. For instance, let q be a smooth real valued function on the circle and let M_q be the operator "multiplication by q." M_q is a bounded self-adjoint operator on $L^2(S^1)$, but $f(M_q)$ is hardly ever of trace class. The theorem of Szegö says:

THEOREM. *Let* π_k *be the orthogonal projection of* $L^2(S^1)$ *onto the space of trigonometric polynomials of degree* $\leq k$ *and let* μ^k *be the spectral measure of* $\pi_k M_q \pi_k$. *Then* $\mu^k/(2k+1)$ *tends weakly to a limiting measure as* k *tends to infinity, this limiting measure being*

$$\mu(f) = \frac{1}{2\pi} \int_0^{2\pi} f(q(\theta))\, d\theta\,, \quad \text{for} \quad f \in C(\mathbf{R})\,.$$

Recently Widom has found an ingeneous proof of this theorem which admits wide-ranging generalizations. I'll take up one of these in lecture 3.

(*)In §2 I've unfortunately used the term "spectral function" for the family of spectral projections associated with a self-adjoint operator on Hilbert space.

I have attempted to make the first three lectures in this series elementary enough to be understood by anyone who has a smattering of knowledge of pseudo-differential operators. (The third lecture assumes a few facts about Fourier integral operators, but these are stated in a down-to-earth way.) The fourth lecture is a lot more technical. Boutet de Monvel and I have recently shown that many results concerning the spectral properties of differential operators have analogues for Toeplitz operators. For instance the Weyl-type formula of Hörmander alluded to in lecture 1 has a Toeplitz analogue, as does the generalized Poisson summation formula of Chazarain [7] and Duistermaat-Guillemin [10] and the clustering theorem of Helton [13]. In lecture 4 I give a couple of applications of this Toeplitz spectral theory to complex analysis. The Szegö-type theorems discussed in this section are Toeplitz analogues of the recent Szegö-type theorems of Weinstein, Widom, Spencer and Kac for the Schroedinger operator. The Weinstein-Widom-Spencer-Kac theorems concern differential operators with periodic bicharacteristics. It turns out that differential operators with this property are rare, whereas quite a lot of interesting Toeplitz operators have this property; so the Toeplitz setting is in some sense more packed with potential. The two representative Szegö theorems described in §4 have the virtue of being fairly elementary to state (though the proofs are technical); so hopefully parts of this lecture will be as accessible to a general audience as the first three were intended to be.

I would like to thank Lars Hörmander, Bob Seeley, and Harold Widom for their advice and assistance.

1. *The lattice point problem*

I will begin with a simple example of the type of formula I want to consider in these lectures. Let D be a compact smooth convex region in $\mathbf{R}^n$ with the origin in its interior and let Z^n be the integer lattice. Let $N^{\#}(\lambda)$ be the cardinality of $\lambda D \cap Z^n$. Then

$$N^{\#}(\lambda) = \lambda^n \operatorname{vol}(D) + O(\lambda^{n-1}) . \tag{1.1}$$

The proof of this is completely elementary. We can estimate the volume of D from below by the number of points of the lattice $\frac{1}{\lambda} Z^n$ which are contained in $(1 - \frac{C}{\lambda})D$ and from above by the number of points which are contained in $(1 + \frac{C}{\lambda})D$. The discrepancy between the two estimates is less than $\frac{C'}{\lambda}$ times the volume of ∂D.

Let D be defined by the inequality

$$D = \{\xi, p(\xi) \leq 1\}$$

where p is a smooth, positive, homogeneous function of degree one on $\mathbf{R}^n - 0$. Let $\chi(\xi)$ be a smooth function on $\mathbf{R}^n$ which is 0 near the origin and 1 for $|\xi| \geq 1$, and let P_0 be the constant-coefficient pseudo-differential operator

$$f \to \frac{1}{(2\pi)^n} \int \chi(\xi)\, p(\xi)\, e^{ix\cdot\xi} \hat{f}(\xi)\, d\xi \,.$$

Since P_0 commutes with translations it maps periodic functions into periodic functions, so it defines a positive self-adjoint pseudo-differential operator $P: C^\infty(T^n) \to C^\infty(T^n)$ the eigenvalues of which are

$$\{p(2\pi z),\ z \in Z^n\}\,.$$

Let $N(\lambda)$ be the number of eigenvalues less than or equal to λ. Then from (1.1) we obtain

$$N(\lambda) = \frac{\lambda^n}{(2\pi)^n} \operatorname{vol}(D) + O(\lambda^{n-1})\,. \tag{1.2}$$

This result is a very special case of the following general result of Hörmander:

THEOREM. *Let* X *be a compact, boundary-less differentiable manifold and let* $P: C^\infty(X) \to C^\infty(X)$ *be a positive self-adjoint first order elliptic pseudo-differential operator. Let* $N(\lambda)$ *be the number of eigenvalues of* P *less than or equal to* λ. *Then*

(1.3) $$N(\lambda) = \frac{\lambda^n}{(2\pi)^n}\,\mathrm{vol}\,(B^*X) + O(\lambda^{n-1})$$

where B^*X *is the subset of the cotangent bundle of* X *on which the symbol of* P *is* ≤ 1.

This theorem and variants of it are going to preoccupy us for the next three lectures. For the moment I want to go back to the simple example (1.1) and consider the question of to what extent the error term in (1.1) is best possible.

Suppose first of all that the domain D in (1.1) has the following property. Suppose there exists a point $\xi_0 \in \partial D$ such that the normal direction to ∂D at ξ_0 is rational and such that the tangent plane to ∂D at ξ intersects ∂D in a neighborhood of ξ_0. (See figure 1.) Letting H be this tangent plane, it is clear that some homothety, $\lambda_0 H$, contains an integer point. Then the number of integer points in λD jumps by an amount of order $O(\lambda^{n-1})$ everytime λ/λ_0 takes on an integer value. Therefore, for such a D, the error term $O(\lambda^{n-1})$ in (1.1) is optimal. More generally, if D is very flat at ξ_0, the error term is close to being optimal.

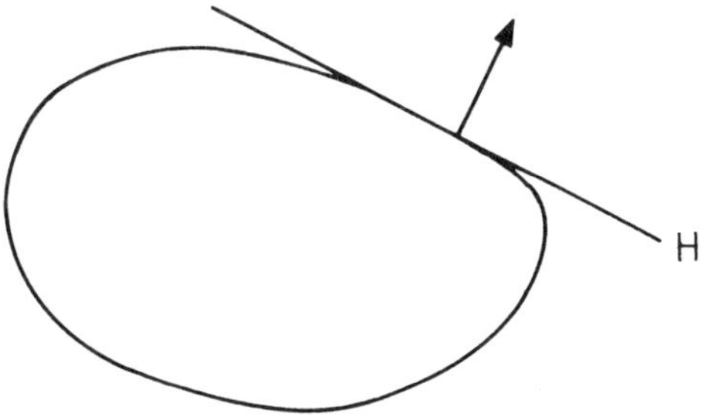

Fig. 1

THEOREM. *Let* ξ_0 *be a point on* ∂D *such that the normal to* ∂D *at* ξ_0 *is rational and the curvature of* ∂D *vanishes to* k-th *order,* $k > n-3$, *then the error term in* (1.1) *is at least of order* $(n-1)(1-1/(k-n+3))$.

Proof. Without loss of generality we can assume the tangent plane, H, to ∂D at ξ_0 is the plane, $\xi_n = 1$. Consider the disk of radius ε on H

centered at ξ_0. Let V_ε be the part of ∂D lying below this disk. (See figure 2.) Each point of V_ε is a distance of order ε^{k+2} from H, so for some constant, c, not depending on ε, the homothety

$$\xi \to (1+c\varepsilon^{k+2})\xi$$

maps V_ε onto a region lying above H. Therefore, if λ is an integer, there are at least $O(\varepsilon\lambda)^{n-1}$ eigenvalues on the interval between λ and $\lambda(1+c\varepsilon^{k+2})$. Now set $\varepsilon = \lambda^{-a/(k+2)}$ where $a > (k+2)/(k-n+3) > 1$. This interval is of length

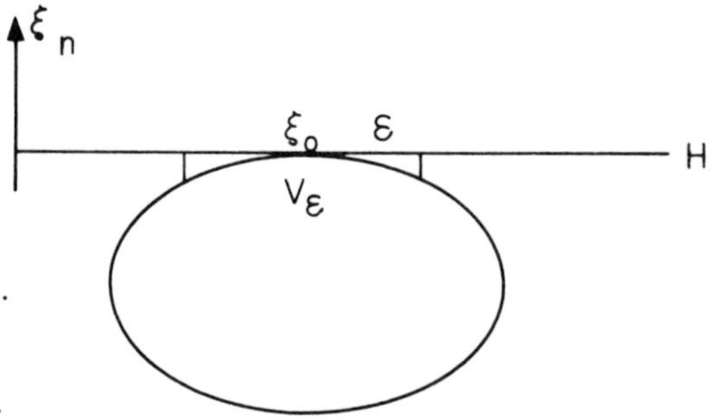

Fig. 2

$c\lambda^{-a+1}$, so if the error term in (1.2) is of order $O(\lambda^\beta)$ the number of eigenvalues on this interval is of order

$$cn\,\frac{\mathrm{vol}\,(D)}{(2\pi)^n}\,\lambda^{n-a} + O(\lambda^\beta)\,. \tag{1.4}$$

On the other hand, since $(\varepsilon\lambda)^{n-1} = \lambda^{(n-1)(1-a/(k+2))}$, (1.4) must be at least of order $O(\lambda^{(n-1)(1-a/k+2)})$, and it is easy to see that if a is chosen as above, that is, $a > (k+2)/(k-n+3)$, then the first term in (1.4) is of order less than this. Thus $\beta \geq (n-1)(1-\frac{a}{k+2})$ for all $a > (k+2)/(k-n+3)$. In particular $\beta \geq (n-1)(1-\frac{1}{k-n+3})$. Q.E.D.

At the other extreme, one has the following estimate:

THEOREM. *If* D *is strictly convex then the error term in* (1.2) *is of order* $O(\lambda^{n-2+2/(n+1)})$.

(This theorem is due to Van der Corput [31] in dimension 2 and to Hlawka [15] and Herz [14] in higher dimensions. See also [19].)

Proof. We will use the Poisson summation formula to write the error term in a form which can be estimated by stationary phase. Let $\chi(\xi)$ be the characteristic function of D and $\chi_\lambda(\xi) = \chi(\frac{\xi}{\lambda})$, the characteristic function of λD. Then

$$N^{\#}(\lambda) = \sum_{\nu \in Z^n} \chi_\lambda(\nu) .$$

To apply the Poisson summation formula to the left hand side, we must first of all regularize χ_λ. To do so, let ρ be a positive smooth function on R^n with support in the unit ball and integral one. Let $\rho_\varepsilon(\xi) = \frac{1}{\varepsilon^n}\rho(\frac{\xi}{\varepsilon})$ and let

(1.5)
$$N_\varepsilon^{\#}(\lambda) = \sum_{\nu \in Z^n} (\chi_\lambda * \rho_\varepsilon)(\nu) .$$

If ν lies a distance greater than ε from the boundary of λD, $\chi_\lambda * \rho_\varepsilon(\nu) = \chi_\lambda(\nu)$; so $N_\varepsilon^{\#}(\lambda - C\varepsilon) \le N^{\#}(\lambda) \le N_\varepsilon^{\#}(\lambda + C\varepsilon)$. Therefore if $N_\varepsilon^{\#}$ satisfies Van der Corput's estimate:

$$(\lambda-\varepsilon)^n \operatorname{vol}(D) + O(\lambda^{n-2+2/n+1}) \le N^{\#}(\lambda) \le (\lambda+\varepsilon)^n \operatorname{vol}(D) + O(\lambda^{n-2+2/n+1}) .$$

Let $\varepsilon = \lambda^{-1+2/n+1}$. Then $(\lambda+\varepsilon)^n = \lambda^n + O(\lambda^{n-2+2/n+1})$; hence $N^{\#}$ satisfies Van der Corput's estimate. Applying the Poisson summation formula to (1.5) we get

$$N_\varepsilon^{\#}(\lambda) = \sum_{\nu \in Z^n} \lambda^n \hat{\chi}(2\pi\lambda\nu)\hat{\rho}(2\pi\varepsilon\nu) .$$

Now $\hat{\chi}(0) = \operatorname{vol}(D)$ and $\hat{\rho}(0) = 0$; so the error term in (1.1) is estimable by

(1.6)
$$\sum_{\nu \in Z^n - 0} \lambda^n \hat{\chi}(2\pi\lambda\nu)\,\hat{\rho}(2\pi\varepsilon\nu) .$$

In a second we will show that for every compact $K \subset\subset R^n - 0$, one has the estimate

$$|\hat{\chi}(\tau x)| = O(\tau^{-(n+1)/2}) \tag{1.7}$$

uniformly for $x \in K$. Assuming this we can estimate (1.6) by

$$\lambda^{n-(n+1)/2} \sum_{\nu \neq 0} |\nu|^{-(n+1)/2} (1+|\varepsilon\nu|^2)^{-N}, \quad \text{with } N \text{ large},$$

or alternatively by the integral

$$\lambda^{(n-1)/2} \int \frac{1}{|x|^{(n+1)/2}} (1+|\varepsilon x|^2)^{-N}\, dx\,.$$

Making the substitution $y = \varepsilon x$, this becomes

$$\lambda^{(n-1)/2} \varepsilon^{-(n-1)/2} \int \frac{1}{|y|^{(n+1)/2}} (1+|y|^2)^{-N}\, dy = O(\lambda^{n-2+2/(n+1)})\,,$$

which is what we want. To establish the estimate (1.7) we write

$$\hat{\chi}(\tau x) = \int_D e^{i\tau x\cdot y}\, dy = \frac{i}{\tau} \int_{\partial D} e^{i\tau x\cdot y}\, du$$

where $du = \frac{1}{|x|^2} \sum (-1)^i x_i\, dy_1 \wedge \cdots \wedge \widehat{dy_i} \wedge \cdots \wedge dy_n$. Now the second integral above is just an oscillatory integral with the phase function $x\cdot y \restriction \partial D$. The assumption that D is strictly convex insures that this phase function has only non-degenerate critical points on ∂D. Therefore, by the lemma of stationary phase the integral is estimable by $O(\tau^{-(n-1)/2})$ the estimate being uniform in x. Q.E.D.

REMARKS. In dimension 2 Van der Corput has shown that this result is optimal. (See [19].) Recently Colin de Verdiere [8] has shown that one can obtain the result above without the assumption of strict convexity. However, one has to assume very strong irrationality conditions on the normal directions of the points where the curvature of the boundary vanishes.

Colin has also obtained the following sharp results in dimension 2. (See also Randol, [26].)

THEOREM. *Suppose* k *is the maximum order for which the curvature of the boundary vanishes. Then if* $k \geq 1$ *the error estimate in* (1.1) *is* $O(\lambda^{1-(1/k+2)})$. *If, at some point at which the curvature vanishes to order* k *the normal direction is rational, then this result is best possible.*

(Note that the "best possible" is a one order improvement on the result we obtained above by our crude geometric argument.)

For the non-constant coefficient case, not much is known at present about the optimality of the error term in Hörmander's estimate (1.3). I'll content myself here with describing a few weak results which are in somewhat the same spirit as those above. As in (1.3) let X be a compact boundary-less manifold and P a positive, self-adjoint elliptic pseudo-differential operator of order one. Let p be the symbol of P and H_p the Hamiltonian vector field on the cosphere bundle of X associated with p.

THEOREM. *Suppose there exists a* $T > 0$ *and an open subset* U *of the cosphere bundle such that* $(\exp TH_p)(x,\xi) = (x,\xi)$ *for all* $(x,\xi) \in U$. *Then the estimate* $O(\lambda^{n-1})$ *in* (1.3) *is best-possible.*

This result is due to Duistermaat and me. I won't prove it here but let me at least indicate what relation it bears to the results I've described in the constant coefficient case. In the constant coefficient case the Hamiltonian H_p is

$$\sum \frac{\partial p}{\partial \xi_i}(\xi) \frac{\partial}{\partial x_i},$$

so $\operatorname{ext} t H_p$ is the map

$$(x,\xi) \to \left(x + t\frac{\partial p}{\partial \xi}, \xi\right).$$

The trajectory through (x_0, ξ_0) is closed in T^n if and only if $\frac{\partial p}{\partial \xi}(\xi_0)$ points in a rational direction. Let T be the period of this trajectory. Then $\exp T H_p$ makes k-th order contact with the identity mapping at (x_0, ξ_0) if and only if $\frac{\partial p}{\partial \xi} - \frac{\partial p}{\partial \xi}(\xi_0)$ vanishes to k-th order at ξ_0. By a homothety we can assume $\xi_0 \in \partial D$. Since $\frac{\partial p}{\partial \xi}(\xi_0)$ is the normal to ∂D at ξ_0 the condition that the trajectory through (x_0, ξ_0) be closed is equivalent to the condition that the normal direction to ∂D at ξ_0 be rational, and the condition that $\exp T H_p$ make k-th order contact with the identity at (x_0, ξ_0) is equivalent to the condition that the curvature of ∂D vanish to order k at ξ_0. In particular the assumption that $\exp T H_p$ be equal to the identity on an open subset of the cosphere bundle means just that a piece of ∂D is flat and that the normal along this flat piece is rational.

I'll conclude with a couple of (very weak) results about the error term in (1.3) in the general case.

THEOREM (Duistermaat-Guillemin [10]). *Suppose that for all* $T > 0$ *the fixed point set of* $\exp T H_p$ *is of measure zero. Then the error term,* $O(\lambda^{n-1})$, *in* (1.3) *can be replaced by* $o(\lambda^{n-1})$.

THEOREM (Bérard [3]). *Let* X *be a compact, boundary-less, Riemannian manifold with everywhere non-positive sectional curvature and let* $P = \sqrt{-\Delta}$. *Then the error term,* $O(\lambda^{n-1})$, *in* (1.3) *can be replaced by* $O(\lambda^{n-1}/\log \lambda)$.

2. *Weyl-type formulas*

Let Δ be the standard Laplacian on $\mathbf{R}^n$, i.e. $\Delta = -\partial^2/\partial^2 x_1 - \cdots - \partial^2/\partial x_n^2$, and let X be a compact region in $\mathbf{R}^n$ with a smooth boundary. Let $\lambda_0 < \lambda_1 \leq \lambda_2 \leq \cdots$ be the spectrum of the self-adjoint boundary problem

$$\Delta u = \lambda u, \ u = 0 \ \text{ on } \partial X .$$

The classical formula of Weyl-Courant says that if $N(\lambda)$ is the number of $\lambda_i \leq \lambda^2$:

$$(2.1) \quad N(\lambda) = \frac{\gamma_n}{(2\pi)^n} \text{volume } X\lambda^n + O(\lambda^{n-1} \operatorname{Log} \lambda), \ \gamma_n = \text{volume } B^n .$$

Recently Agmon-Hörmander, Brüning [6], and Seeley [28] have extended this formula to arbitrary Riemannian manifolds with boundary. Concerning the error term in (2.1), it is conjectured, and in some instances proved, that the better error term, $O(\lambda^{n-1})$, is valid. The best result I know about along these lines is due to Seeley [28] for domains in R^3. A more audacious conjecture is that the error term is a universal constant times

$$\text{volume } \partial X \lambda^{n-1} + o(\lambda^{n-1}) .$$

As far as I know this hasn't been verified except in very simple cases.

The classical proof of (2.1) uses max-min. techniques and polygonal domains. Today I plan to give a different proof of (2.1) which is due to Seeley and is based on Hörmander's proof of the Weyl formula in the boundary-less case.(*) At the end of my talk I'll say a few words about the generalizations of (2.1) which I mentioned above.

Let $\phi_0, \phi_1, \phi_2, \cdots$ be normalized eigenfunctions corresponding to the eigenvalues $\lambda_0, \lambda_1, \lambda_2, \cdots$. Let $E(\lambda)$ be orthogonal projection on the space spanned by the ϕ_i's with $\lambda_i \leq \lambda^2$. The Schwartz kernel of $E(\lambda)$ is the *spectral function* of X and is given explicitly in terms of the ϕ_i's:

$$(2.2) \qquad e(x, y, \lambda) = \sum_{\lambda_i \leq \lambda^2} \phi_i(x)\phi_i(y) .$$

Obviously $N(\lambda) = \operatorname{trace} E(\lambda) = \int e(x, x, \lambda)\, d\lambda$. It is very hard to get at the asymptotic properties of $e(\lambda)$ directly. What one usually does is take

(*)Hormander has pointed out to me that in the second order case the estimates of the spectral function used in this proof go back to papers of Avakumovic and Levitan from the early 50's. See [20, 21, 22].

some operator transform of $e(x, y, \lambda)$ whose asymptotic properties can be determined by PDE techniques and then use Tauberian theorems to deduce asymptotic properties of e itself. For example, from asymptotic properties of the resolvant kernel

$$\text{S.K.}\ (\Delta - z)^{-1} = \int_0^\infty \frac{1}{\lambda^2 - z} \frac{d}{d\lambda} e(x, y, \lambda)\, d\lambda$$

one can easily deduce the "order of magnitude" estimate

$$e(x, x, \lambda) \leq C\lambda^n, \quad \forall x \in X \tag{2.3}$$

by a Tauberian argument. (See Agmon [1, Theorem 5.11].) An elementary proof of (2.3) for the second order case will be given in an appendix. Notice that by integrating (2.3) we get the bound $N(\lambda) \leqq C\lambda^n$.

Following Hörmander we will work with the cosine transform

$$u(x, y, t) = \text{S.K.}\ \cos t\sqrt{\Delta} = \int_0^\infty \cos \lambda t \frac{d}{d\lambda} e(x, y, \lambda)\, d\lambda \tag{2.4}$$

which is the fundamental solution of the wave equation:

$$-\frac{\partial}{\partial t^2} u(x, y, t) = \Delta_x u(x, y, t),\ u(x, y, t) = 0 \ \text{ for } \ x \in \partial X \tag{2.5}$$

with initial data, $u(x, y, 0) = \delta(x-y)$ and $\dot{u}(x, y, 0) = 0$. One reason for singling out the cosine transform is that there is a very simple Tauberian theorem relating the behavior of $u(x, x, t)$ for small t to the behavior of $e(x, x, \omega)$ for large λ. (In fact this theorem will come up later on in the lecture.) As far as Hörmander was concerned, another compelling reason for singling out the cosine transform is that in the boundaryless case one can write down a very explicit formula for $u(x, y, t)$ when t is small. Unfortunately, if there is a boundary, this is no longer true, and we'll see why in a minute. For a time this deterred people from attempting to adapt the Hörmander approach to the boundary case.

One can construct a fundamental solution of (2.5) as follows. One starts out with the free space solution $u^0(x,y,t)$ of (2.5) on all of R^n. Since this doesn't satisfy the boundary condition one modifies it by adding a "reflection term." For instance if the boundary were just $x_n = 0$ then setting

$$u^0_{\text{reflected}}(x,y,t) = u^0(x,y_1,\cdots,y_{n-1},-y_n,t),$$

the difference $u^0 - u^0_{\text{reflected}}$ satisfies (2.5) in $x_n \geq 0$ with the correct initial and boundary data. If the boundary is compact then even for small values of t one has to add *countably many* such reflected terms to $u^0(x,y,t)$ in order to get the fundamental solution of (2.5). The picture below shows why this is the case:

2 reflections 4 reflections 8 reflections etc.

Seeley gets out of this mess by exploiting the following simple theorem.

THEOREM. *If* y *is at distance greater than* t *from the boundary then*

$$u(x,y,t) = u^0(x,y,t).$$

Proof. Consider $u^0(x,y,t)$ as a function of x with y fixed. The impulse represented by $\delta_y = \delta(x-y)$ takes a time $t = d(y,\partial X)$ to propagate to the boundary; so if t is less than $d(y,\partial X)$ then $u^0(x,y,t) = 0$ on ∂X; i.e. $u^0(x,y,t)$ satisfies the boundary condition.

It is easy to write down a closed form expression for $u^0(x,y,t)$. Letting Δ_0 be the Laplace operator operating on the Schwartz space we have $\widehat{\Delta_0 f} = |\xi|^2 f$ for $f \in \mathcal{S}$. Therefore

(2.6) $$\widehat{\cos t\sqrt{\Delta_0} f} = \cos t|\xi|\hat{f}.$$

Since $u^0(x, y, t)$ is the Schwartz kernel of the operator $\cos t\sqrt{\Delta_0}$ we get from (2.6)

$$u^0(x, y, t) = \frac{1}{(2\pi)^n}\int e^{i(x-y)\cdot\xi}\cos t|\xi|\,d\xi\,.$$

In particular setting $x = y$ we have

$$u^0(x, x, t) = \frac{\omega_n}{(2\pi)^n}\int_0^\infty \cos st\, s^{n-1}\,ds$$

where $\omega_n = n\gamma_n = \mathrm{vol}(S^{n-1})$. Combining this with the previous theorem we get:

$$u(x, x, t) = \frac{\gamma_n}{(2\pi)^n}\int_0^\infty \cos st\,\frac{d}{ds}\,s^n\,ds \tag{2.7}$$

for $d(x, \partial X) > t$ where $u(x, y, t)$ is the fundamental solution of (2.5). To convert (2.7) into a statement about the asymptotic properties of $e(x, x, \lambda)$ we will make use of the following simple Tauberian theorem (cf. Levitan [23]):

THEOREM. *Let $\gamma_n : [0, \infty) \to \mathbf{R}$ be the function $\gamma_n(\lambda) = \lambda^n$ and let $m : [0, \infty) \to \mathbf{R}$ be any non-decreasing function of polynomial growth with $m(0) = 0$. Suppose the cosine transforms of $\frac{dm}{d\lambda}$ and $\frac{d\gamma_n}{d\lambda}$ are equal on the interval, $|t| \leq \delta$. Then*

$$|m(\lambda) - \lambda^n| \leq C_n\left(\frac{1}{\delta^n} + \frac{\lambda^{n-1}}{\delta}\right) \quad \textit{for} \quad \lambda > 0 \tag{2.8}$$

where C_n is a universal constant depending only on n.

We will prove this theorem below. First let's use this result to derive the Weyl-Courant inequality (2.1). Applying the theorem to the expression (2.7) we obtain

(2.9) $$\left| e(x,x,\lambda) - \frac{\gamma_n}{(2\pi)^n}\lambda^n \right| \le C_n\left(\frac{1}{\delta^n} + \frac{\lambda^{n-1}}{\delta}\right)$$

for $d(x,\partial X) > \delta$. Now we can estimate $N(\lambda) - \frac{\gamma_n}{(2\pi)^n}\,\mathrm{vol}(X)\lambda^n$ by

$$\int_X \left| e(x,x,\lambda) - \frac{\gamma_n}{(2\pi)^n}\lambda^n \right| dx\ .$$

We will break this integral up into the integral over the region $d(x,\partial X) \le \frac{1}{\lambda}$ plus the integral over the region $d(x,\partial X) > \frac{1}{\lambda}$. By (2.3), $e(x,x,\lambda) \le C\lambda^n$ uniformly on X; so the first integral is of order $O(\lambda^{n-1})$. By (2.9) the second integral is majorized by a constant times

(2.10) $$\int_{1/\lambda}^{a} \left(\frac{1}{\delta^n} + \frac{\lambda^{n-1}}{\delta}\right) d\delta$$

for a sufficiently large, and (2.10) is clearly of order $O(\lambda^{n-1}\,\mathrm{Log}\,\lambda)$. This concludes the proof of (2.1).

There are a couple of observations one should make about this proof:

1) Suppose one could get an estimate for

$$u(x,x,t) - \frac{\gamma_n}{(2\pi)^n}\int_0^\infty \cos st\,\frac{d}{ds}\,s^n\,ds$$

on an interval that is a little larger than $|t| < d(x,\partial X)$, for instance the interval $|t| < \sqrt{d(x,X)}$. Presumably, by a Tauberian argument, one could convert this into an estimate of the form (2.9) with $\delta = \sqrt{d(x,\partial X)}$. We would then be able to estimate the integral over $d(x,\partial X) = s > \frac{1}{\lambda}$ in the argument above by

$$\int_{1/\lambda}^{a} \left(\frac{1}{s^{n/2}} + \frac{\lambda^{n-1}}{s^{1/2}}\right) ds$$

which is $O(\lambda^{n-1})$. The integral over $d(x, \partial X) < \frac{1}{\lambda}$ being estimable, as before, by $O(\lambda^{n-1})$ we would get the optimal error term in (2.1). Seeley has been able to carry out the details of this argument in $\mathbf{R}^3$, and only minor technical difficulties seem to stand in the way of carrying it out in general. The main point of Seeley's proof is that in a time interval $t = O(\sqrt{d(x, \partial X)})$ a light ray starting from x can make at most *one* reflection in the boundary; so only one reflected term has to be added to $u^0(x, y, t)$ to get the correct fundamental solution of (2.5).

2) With small changes the argument above gives the Weyl formula with the $O(\lambda^{n-1} \operatorname{Log} \lambda)$ error term for Riemannian manifolds with boundary. One needs only to replace $u^0(x, y, t)$ by the fundamental solution of (2.5) on an open manifold containing X, as in Hörmander [16], and to jazz up a little the Tauberian theorem which we used above.

Let me finish this lecture with a quick proof of this Tauberian theorem. First let's extend the function m to the whole real line by setting $m^{\#}(s) = m(|s|) \operatorname{sgn} s$. Similarly let's set $\gamma_n^{\#}(s) = \gamma_n(|s|) \operatorname{sgn} s$. Then the cosine transforms of $\frac{d}{ds} m(s)$ and $\frac{d}{ds} \gamma_n(s)$ are half the Fourier transforms of $\frac{d}{ds} m^{\#}$ and $\frac{d}{ds} \gamma_n^{\#}$; so in the theorem we can replace cosine transforms by Fourier transforms and m and γ by $m^{\#}$ and $\gamma^{\#}$. It is also easy to see, by a change of scale, that we can take $\delta = 1$. We are therefore reduced to proving

THEOREM. *Let* $\gamma_n = \gamma(s)$ *be the function* $\gamma_n(s) = |s|^n \operatorname{sgn} s$ *and let* $m(s)$ *be a non-decreasing function on the real line having polynomial growth, with* $m(0) = 0$. *Suppose the Fourier transforms of* $\frac{dm}{ds}$ *and* $\frac{d}{ds}\gamma_n$ *are equal on the interval,* $|t| < 1$. *Then, for* $s > 0$,

$$|m(s) - s^n| < C_n(1 + s^{n-1})$$

where C_n *is a universal constant depending only on* n.

Proof. We will need a function $\rho \in \mathcal{S}(\mathbf{R}^1)$ with the following properties:

$$(2.11)\qquad \begin{aligned} &\text{i) } \rho(s) = \rho(-s),\ \rho \geq 0 \quad \text{and} \quad \rho > 0 \quad \text{on} \quad [-1,1] \\ &\text{ii) } \hat{\rho}(0) = 1,\ \hat{\rho} \leq 1 \quad \text{and} \quad \hat{\rho} = 0 \quad \text{for} \quad |t| > 1\,. \end{aligned}$$

It is easy to construct such a function. Let ϕ be a smooth function on $\mathbf{R}^1$ which is positive, even and supported in $[-\frac{1}{2}, \frac{1}{2}]$. Let ρ_0 be defined by

$$\hat{\rho}_0 = \frac{1}{c}\phi * \phi \quad \text{where} \quad c = (\phi * \phi)(0)\,.$$

Then $\rho_0 \geq 0$, $\rho_0(s) = \rho_0(-s)$, $\hat{\rho}_0(0) = 1$ and $\hat{\rho}_0(t) = 0$ for $|t| > 1$.

$$\hat{\rho}_0(t) \leq \int |\rho_0|\,ds = \int \rho_0\,ds = 1\,.$$

Since $\rho_0(0) = \frac{1}{2\pi}\int \hat{\rho}_0(t)\,dt > 0$, $\rho_0(s) > 0$ on a small interval $|s| < a < 1$. Setting $\rho(s) = a\rho_0(as)$, this will do the trick.

In what follows C_n will denote *some* universal constant depending only on n. We will, however, allow this constant to change from line to line.

LEMMA 1. $|(\rho * \gamma_n - \gamma_n)(s)| \leq C_n(1+|s|^{n-1})$.

Proof. $\rho * \gamma_n(s) = \int_{-\infty}^{\infty} \rho(r)|s-r|^n\,dr = |s|^n \int_{-\infty}^{\infty} \rho(r)\,dr + \cdots = |s|^n + \cdots$

where the dots are estimable by lower powers of $|s|$. Q.E.D.

LEMMA 2. $|\rho * \frac{dm}{ds} - n|s|^{n-1}| \leq C_n(1+|s|^{n-2})$.

Proof. Since $\hat{\rho}$ is supported in the interval $|t| < 1$, $\rho\widehat{\frac{dm}{ds}} = \rho\widehat{\frac{d\gamma_n}{ds}}$; so $\rho * \frac{dm}{ds} = \rho * \frac{d\gamma_n}{ds}$. By Lemma 1, $|\rho * \frac{d\gamma_n}{ds} - \frac{d\gamma_n}{ds}| \leq C_n(1+|s|^{n-2})$; so this implies the above estimate. Q.E.D.

LEMMA 3. $|m(s+1)-m(s)| \leq C_n(1+|s|^{n-1})$.

Proof. By (2.11) $\rho \geq 0$ and $\rho(s) > c > 0$ for $|s| \leq 1$. Since m(s) is monotone

$$|m(s+1)-m(s)| = \int_s^{s+1} \frac{dm}{dr}(r)\,dr \leq \frac{1}{c}\int_s^{s+1} \rho(s-r)\frac{dm}{dr}\,dr \leq \frac{1}{c}\rho * \frac{dm}{dr},$$

so Lemma 3 follows from Lemma 2. Q.E.D.

LEMMA 4. *For all* σ

$$|m(s+\sigma)-m(s)| \leq C_n(1+|\sigma|^n)(1+|s|^{n-1}).$$

Proof. It's enough to prove this with σ a positive integer. We can write

$$\begin{aligned} m(s+\sigma)-m(s) &= m(s+1)-m(s)+m(s+2)-m(s+1)+\cdots \quad (\sigma \text{ terms}) \\ &\leq C_n|\sigma|(1+(|s|+\sigma)^{n-1}). \end{aligned}$$

With another C_n we get the estimate above. Q.E.D.

We'll now prove this theorem: By Lemma 4

$$|\rho * m(s)-m(s)| \leq \int \rho(-\sigma)(|m(s+\sigma)-m(s)|)\,d\sigma \leq C_n(1+|s|^{n-1})$$

since ρ is rapidly decreasing. Integrating the estimate in Lemma 2 we get

$$|(\rho * m)(s)-|s|^n| \leq C_n(1+|s|^{n-1}).$$

and combining these two estimates gives us the theorem.

REMARK. This Tauberian theorem is a somewhat simple-minded version of the Tauberian theorem used by Hörmander in [16].

Appendix

In the proof of (2.1) we made use of the order of magnitude estimate for the spectral function

$$e(x, x, \lambda) \leq C\lambda^n , \tag{A.1}$$

C being independent of x. Agmon has proved very general estimates of the form (A.1) using the Stieljes transform of the spectral function. It was pointed out to us by Hörmander that, for the problem considered in this lecture, namely the Dirichlet problem for the Laplace operator on a bounded domain, X, in R^n, there is a very elementary proof of (A.1) based on the minimum principle for the heat equation. We will give the proof suggested by Hörmander in this appendix.

Consider the Laplace transform of the spectral function

$$v(x, y, t) = \int_0^\infty e^{-\lambda^2 t} \frac{d}{d\lambda} e(x, y, \lambda)\, d\lambda . \tag{A.2}$$

It is clear that for fixed $y \in X$, $v(x, y, t)$ is a solution of the heat equatior

$$\frac{\partial}{\partial t} v + \Delta v = 0$$

with initial data $v(x, y, 0) = \delta(x-y)$ and boundary data, $v(x, y, t) = 0$ when $x \in \partial X$. Let $v_0(x, y, t)$ be the free space solution of the heat equation:

$$v_0(x, y, t) = \left(\frac{1}{4\pi t}\right)^{n/2} e^{-(x-y)^2/4t} . \tag{A.3}$$

Fix $y \in$ Interior X and consider $w(x, t) = v_0(x, y, t) - v(x, y, t)$ on the region

$$\{(x, t),\ x \in X\ \ 0 \leq t < \infty\} . \tag{A.4}$$

$w(x, t)$ is zero for $t = 0$ since v and v_0 have the same initial conditions, and $w(x, t)$ is non-negative on the boundary since v_0 is everywhere non-negative and v vanishes on the boundary. By the minimum principle for the heat equation,(*) $w(x, t) \geq 0$ on the entire region (A.4).

(*)See for instance Petrowsky [25].

Setting $x = y$ we get

$$v(x, x, t) \leq \left(\frac{1}{4\pi t}\right)^{n/2} \tag{A.5}$$

for all x in X. We now appeal to a rather trivial Tauberian theorem for the Laplace transform.

THEOREM. *Let $\alpha(\lambda)$, $0 \leq \lambda < \infty$, be a non-decreasing piece-wise continuous function with $\alpha(0) = 0$. If*

$$\int_0^\infty e^{-\lambda^2 t} \frac{d}{d\lambda} \alpha(\lambda)\, d\lambda \leq C\, t^{-n/2}$$

for all t, then $\alpha(\lambda) \leq (eC)\lambda^n$ for all λ.

Proof. For all $\lambda_0 > 0$ and all $t > 0$

$$e^{-\lambda_0^2 t} \alpha(\lambda_0) = \int_0^{\lambda_0} e^{-\lambda_0^2 t} \frac{d}{d\lambda} \alpha(\lambda)\, d\lambda \leq \int_0^{\lambda_0} e^{-\lambda^2 t} \frac{d}{d\lambda} \alpha(\lambda)\, d\lambda \leq C t^{-n/2}.$$

Setting $t = 1/\lambda_0^2$, we obtain

$$e^{-1} \alpha(\lambda_0) \leq C \lambda_0^n. \qquad \text{Q.E.D.}$$

Applying this theorem to the inequality (A.5), we get

$$e(x, x, \lambda) \leq C \lambda^n$$

for all $\lambda > 0$, with $C = e(4\pi)^{-n/2}$.

3. *Szegö-type formulas I*

Let A be a densely defined self-adjoint operator on Hilbert space whose spectrum is discrete and has no finite points of accumulation. The *spectral measure* of A is, by definition, the measure

(3.1) $$\sum \delta(\lambda - \lambda_i), \quad \lambda_i \in \text{spec } A .$$

A convenient way of defining this measure is in terms of traces. Let f be a continuous function on the real line with compact support. The spectral theorem permits us to form f(A), and if the spectrum of A has the properties above, f(A) is an operator of finite rank, so, *a fortiori*, of trace class. The map

$$f \in C_0(R) \to \text{trace } f(A)$$

is a continuous functional on $C_0(R)$; therefore, by the Riesz representation theorem, there exists a measure μ_A on the real line such that $\mu_A(f) = \text{trace } f(A)$. The measure μ_A is obviously just the measure (3.1).

If the operator A has continuous spectrum or has discrete spectrum with finite points of accumulation, then (3.1) as it stands doesn't make sense. Sometimes, however, one can define a kind of "renormalized" spectral measure for A in the following way. Let $\pi_1, \pi_2, \pi_3, \cdots$ be a family of orthogonal projection operators on Hilbert space satisfying

(3.2)
$$\begin{aligned} &\text{a) } \operatorname{rank} \pi_i = r_i < \infty , \\ &\text{b) } \pi_i < \pi_j \quad \text{for} \quad i < j , \\ &\text{c) } \lim \pi_i = I \quad \text{as} \quad i \to \infty . \end{aligned}$$

Let μ_i be the spectral measure of $\pi_i A \pi_i$. Then in certain instances the sequence $\{\mu_i / r_i\}$ has a weak limit, μ, as i tends to infinity. If it exists, μ is a plausible substitute for (3.1). (Of course there may be different μ's for different choices of the π_i's.)

The classical Szegö theorem is one of the simplest examples of a result of the above type. Let q be a smooth function on the circle and let $M_q : L^2(S^1) \to L^2(S^1)$ be the operator "multiplication by q." Let $\pi_k : L^2(S^1) \to L^2(S^1)$ be the orthogonal projection of $L^2(S^1)$ onto the space spanned by $\{e^{im\theta}, |m| \le k\}$. The Szegö theorem ([29], [17], [33]) says:

THEOREM. *If* μ_k *is the spectral measure of* $\pi_k M_q \pi_k$ *then* $\mu_k/(2k+1)$ *converges weakly to the measure,* μ, *defined by*

$$\mu(f) = \frac{1}{2\pi}\int f(q(\theta))\,d\theta, \qquad \forall f \in C(R).$$

Widom has recently given a very simple proof of this result which is susceptible to wide-ranging generalizations. I want to discuss one of these generalizations in this lecture.

Let X be a compact n-dimensional manifold and let P be a first order elliptic pseudo-differential operator on X which is self-adjoint and positive. Let p be the symbol of P and let S^*X be the subset $\{(x,\xi),\ p(x,\xi) = 1\}$ of the cotangent bundle of X. S^*X carries a canonical measure, the Liouville measure, ν, defined by

$$\nu = \frac{d}{dt}\chi_t\, dx \wedge d\xi\,\Big|_{t=1}$$

where $dx \wedge d\xi$ is the symplectic measure on T^*X and χ_t the characteristic function of the set, $p \leq t$.

THEOREM 1. *Let* π_λ *be the orthogonal projection of* $L^2(X)$ *onto the space spanned by the eigenfunctions of* P *of eigenvalue* $\leq \lambda$. *Then for every bounded self-adjoint pseudo-differential operator* $B: L^2(X) \to L^2(X)$ *the following is true: if* μ_λ *is the spectral measure of* $\pi_\lambda B \pi_\lambda$, *the sequence of measures* $\{\mu_\lambda/\lambda^n\}$ *converges weakly to the measure* μ *defined by*

$$\mu(f) = \frac{1}{(2\pi)^n n}\int_{S^*X} f(\sigma(B)(x,\xi))\,d\nu, \qquad \forall f \in C(R). \tag{3.3}$$

In particular, if $X = S^1$ and $P = (-d^2/d\theta^2)^{1/2}$, Theorem 1 is Szegö's theorem. The proof of this theorem which we are going to give below is, except for small modifications, due to Widom. Widom proves a much stronger result than Theorem 1 in [37], but for a very special class of P's and X's.

Our first goal is to prove that Theorem 1 is true for the function, $f = x$. Let $\phi(\lambda) = \text{trace}\ \pi_\lambda B \pi_\lambda$ and let $\gamma(t)$ be the Fourier-Stieljes transform

$$\int e^{i\lambda t} \frac{d\phi}{d\lambda}\, d\lambda\ .$$

Let $b(x,\xi) = \sigma(B)(x,\xi)$ and let $\nu(b)$ be the integral

$$\frac{1}{(2\pi)^n} \int_{S^*X} b(x,\xi)\, d\nu\ .$$

LEMMA 1. *For small values of* t,

$$\gamma(t) = \nu(b) \int_0^\infty \lambda^{n-1} e^{i\lambda t}\, d\lambda + \hat{h}(t) \tag{3.4}$$

where $h(\lambda)$ *is of order* $O(\lambda^{n-2})$ *for* λ *large.*

Proof. Let $U(t) = \exp \sqrt{-1}\, tP$. According to Hörmander [1], for $f \in C^\infty(X)$, $U(t)f$ can be expressed as an oscillatory integral of the form

$$\frac{1}{(2\pi)^n} \int (1 + a_{-1}(x,\xi))\, e^{i(tp(y,\xi) + q(x,y,\xi))} f(y)\, dy d\xi \quad \text{mod } C^\infty\ ,$$

for small values of t, where $dyd\xi$ is the symplectic volume on T^*X, q is a homogeneous function of degree 1 in ξ which vanishes when $x = y$ and $a_{-1}(x,\xi)$ is a symbol of type $S^{-1}_{1,0}$. The operator $BU(t)$, applied to f, can be expressed similarly, namely as:

$$\frac{1}{(2\pi)^n} \int b(x,y,\xi)\, e^{i(tp(y,\xi) + q(x,y,\xi))} f(y)\, dy d\xi \quad \text{mod } C^\infty \tag{3.5}$$

where $b(x,x,\xi)$ has the principal part $b(x,\xi)$. If we compute the trace of $BU(t)$ from (3.5) we get, for small values of t,

$$\text{trace } BU(t) = \frac{1}{(2\pi)^n} \int (b(x,\xi) + b_{-1}(x,\xi))\, e^{itp(x,\xi)}\, dx d\xi + r(t)$$

where $r(t) \in C_0^\infty(R)$. Introducing $p(x,\xi) = \lambda$ as a new independent variable in the integrand we can rewrite the integral in the form

$$\frac{1}{(2\pi)^n} \int_0^\infty \lambda^{n-1} \int_{S^*X} e^{it\lambda}\, (b(x,\xi) + b_{-1}(x,\lambda\xi))\, d\nu d\lambda + r(t)$$

or alternatively in the form

$$\nu(b) \int_0^\infty \lambda^{n-1}\, e^{i\lambda t}\, d\lambda + \hat{h}(t)$$

where if s is the inverse Fourier transform of r

$$h(\lambda) = \lambda_+^{n-1} \int_{S^*X} b_{-1}(x,\lambda\xi)\, d\nu + s(\lambda)\,.$$

On the other hand, if $\lambda_1 \le \lambda_2 \le \lambda_3 \le \cdots$ are the eigenvalues of P and $\phi_1, \phi_2, \phi_3, \cdots$ a corresponding sequence of normalized eigenfunctions, then, formally, the Schwartz kernel of the operator $BU(t)$ is

$$\sum B\phi_i(x)\, \bar{\phi}_i(y)\, e^{\sqrt{-1}\,\lambda_i t}$$

and for the trace of $BU(t)$ we get, formally,

$$\sum \langle B\phi_i, \phi_i \rangle\, e^{\sqrt{-1}\,\lambda_i t}\,. \tag{3.6}$$

By definition

$$\phi(\lambda) = \text{trace } \pi_\lambda B \pi_\lambda = \sum_{\lambda_i \le \lambda} \langle B\phi_i, \phi_i \rangle\,, \tag{3.7}$$

so (3.6) is just the Fourier transform of $\frac{d\phi}{d\lambda}$. Q.E.D.

LEMMA 2. *The function* $\phi(\lambda)$ *satisfies*

$$|\phi(\lambda+\sigma)-\phi(\lambda)| \leq C|\sigma|(|\lambda|^{n-1}+|\sigma|^{n-1}) \tag{3.8}$$

with a constant C *independent of* σ *and* λ.

Proof. By Hörmander's spectral theorem (see section 1) there exists a C such that the number of eigenvalues of P on the interval between λ and $\lambda+\sigma$ is $\leq C|\sigma|(|\lambda|^{n-1}+|\sigma|^{n-1})$. Since $B: L^2(X) \to L^2(X)$ is bounded, the terms $\langle B\phi_i, \phi_i \rangle$ in (3.7) are uniformly bounded; so with another C we get the estimate (3.8). Q.E.D.

LEMMA 3. *If* ϕ *is any function on the real line satisfying* (3.8) *and* ρ *is a Schwartz function with* $\int \rho(\sigma)\,d\sigma = 1$, *then* $\phi = \rho * \phi + O(\lambda^{n-1})$.

Proof. We have

$$\begin{aligned}(\rho*\phi-\phi)(\lambda) &= \int \{\phi(\lambda-\sigma)-\phi(\lambda)\}\rho(\sigma)\,d\sigma \\ &\leq C|\lambda|^{n-1}\int |\sigma|\,\rho(\sigma)\,d\sigma + C\int |\sigma|^n \rho(\sigma)\,d\sigma \\ &= O(\lambda^{n-1}).\end{aligned}$$

Q.E.D.

Combining these lemmas we get the following weak form of Theorem 1.

THEOREM 1′. *For every bounded self-adjoint pseudo-differential operator* $B: L^2(X) \to L^2(X)$

$$\text{trace}\,\frac{\pi_\lambda B \pi_\lambda}{\lambda^n} = \frac{1}{(2\pi)^n n}\int_{S^*X} \sigma(B)(x,\xi)\,d\nu + O(1/\lambda). \tag{3.9}$$

Proof. Set $\psi(\lambda) = \nu(b)\,\lambda_+^n/n + H(\lambda)$ where $H(\lambda) = \int_0^\lambda h(s)\,ds$. By Lemma 1 there exists an $\varepsilon > 0$ such that the Fourier transform of $\frac{d}{d\lambda}(\phi-\psi)$ is zero on the interval $-\varepsilon < t < \varepsilon$. Choose $\rho \in \mathcal{S}(\mathbf{R})$ such that $\int \rho(\sigma)\,d\sigma = 1$ and the support of $\hat{\rho}$ is contained in the interval $-\varepsilon < t < \varepsilon$. Then

$$\frac{d}{d\lambda}\rho * \phi = \frac{d}{d\lambda}\rho * \psi .$$

Since $\rho * \phi(-\infty) = \rho * \psi(-\infty) = 0$, this implies that $\rho * \phi = \rho * \psi$. Both ψ and ϕ satisfy the hypotheses of Lemma 3; so $\phi = \psi + O(\lambda^{n-1})$. Q.E.D.

To convert (3.9) into a statement about the spectral measure of $\pi_\lambda B \pi_\lambda$ we resort to an argument of Widom, [37]. Let "$\|\ \|$" be the sup norm, "$\|\ \|_1$" the trace norm, and "$\|\ \|_2$" the Hilbert-Schmidt norm for finite rank operators on $L^2(X)$.

LEMMA 4. *If* $B: L^2(X) \to L^2(X)$ *is a zeroth order pseudo-differential operator,*

$$\|\pi_\lambda B(1-\pi_\lambda)\|_2 = o(\lambda^{n/2}) .$$

Proof. If λ_1 and λ_2 are distinct eigenvalues of P and π_1 and π_2 the orthogonal projections of $L^2(X)$ onto their eigenspaces

$$\pi_1 B \pi_2 = (\lambda_1 - \lambda_2)^{-1} \pi_1 [P, B] \pi_2 . \tag{3.10}$$

Now for any λ and δ, $\delta > 0$,

$$\|\pi_\lambda B(1-\pi_\lambda)\|_2^2 = \|\pi_\lambda B(\pi_{\lambda+\delta} - \pi_\lambda)\|_2^2 + \|\pi_\lambda B(1-\pi_{\lambda+\delta})\|_2^2 .$$

Since $[P, B]$ is bounded, we can use (3.10) to bound the second term on the right by $\delta^{-2}\|\pi_\lambda [P,B]\|_2^2$. Thus $\|\pi_\lambda B(1-\pi_\lambda)\|_2^2$ is bounded from above by

$$\|B\|^2 \text{ rank } (\pi_{\lambda+\delta} - \pi_\lambda) + \delta^{-2}\|[P,B]\|^2 \text{ rank } \pi_\lambda .$$

According to the theorem of Hörmander cited above

$$\text{rank } (\pi_{\lambda+\delta} - \pi_\lambda) \le C|\delta|(|\lambda|^{n-1} + |\delta|^{n-1}) ,$$

and rank $\pi_\lambda \le C\lambda^n$ by Weyl's lemma. Therefore, with another constant C, we get the estimate

$$\lambda^{-n}\|\pi_\lambda B(1-\pi_\lambda)\|^2 \le C\left\{|\delta|\left(\frac{1}{\lambda}+\frac{|\delta|^{n-1}}{\lambda^n}\right)+\frac{1}{\delta^2}\right\}.$$

Fixing δ and letting λ tend to infinity we get

$$\overline{\lim}\,\lambda^{-n}\|\pi_\lambda B(1-\pi_\lambda)\|^2 \le \frac{C}{\delta^2}\ ;$$

however, since δ is arbitrarily large, the lemma follows. Q.E.D.

We now prove Theorem 1. Set $\pi^\perp = 1-\pi$ and consider $\pi_\lambda B^k \pi_\lambda = \pi_\lambda B(\pi_\lambda+\pi_\lambda^\perp)B(\pi_\lambda+\pi_\lambda^\perp)\cdots B\pi_\lambda$. Expanding this we get a sum of products one of which is

$$\pi_\lambda B\pi_\lambda\ B\pi_\lambda\ \cdots\ B\pi_\lambda = (\pi_\lambda B\pi_\lambda)^k$$

while in any other product the factor $\pi_\lambda B\pi_\lambda^\perp$ appears. But, by Lemma 4,

$$\|\pi_\lambda B\pi_\lambda^\perp\|_1 \le \|\pi_\lambda\|_2\,\|\pi_\lambda B\pi_\lambda^\perp\|_2 = o(\lambda^n)\ .$$

Thus

$$\|\pi_\lambda B^k\pi_\lambda - (\pi_\lambda B\pi_\lambda)^k\|_1 = o(\lambda^n)\ .$$

More generally for any polynomial function, f, of B

$$\|\pi_\lambda f(B)\pi_\lambda - f(\pi_\lambda B\pi_\lambda)\|_1 = o(\lambda^n)\ ,$$

so $\lambda^{-n}(\text{trace}\ \pi_\lambda f(B)\pi_\lambda - \mu_\lambda(f)) = o(1)$, where μ_λ is the spectral measure of $\pi_\lambda B\pi_\lambda$. As λ tends to infinity, λ^{-n} trace $\pi_\lambda f(B)\pi_\lambda$ tends to

$$\frac{1}{(2\pi)^n n}\int_{S^*X} f(\sigma(B)(x,\xi))\,d\nu$$

by Theorem 1′; so the same is true for $\lambda^{-n}\mu_\lambda(f)$. This proves Theorem 1 for polynomial functions, f. If g is any continuous function on the real line such that $\sup |g-f| < \varepsilon$ then $\|\mu_\lambda(g)-\mu_\lambda(f)\| \lambda^{-n} \leq \varepsilon C$ with a constant not depending either on λ or ε; so, by the Stone-Weierstrass theorem, Theorem 1 is true for all continuous functions, g.

I'll end this lecture by describing a generalization of the classical Szegö theorem to n dimensions. Again this result is due to Widom ([35]).

Let D be a convex domain in $\mathbf{R}^n$ and let $\lambda D = \{\lambda\xi, \xi \in D\}$. Let T^n be the standard n-torus; $\mathbf{R}^n/2\pi Z^n$. For each $\lambda > 0$ let π_λ be the orthogonal projection of $L^2(T^n)$ onto the space spanned by

$$\{e^{i x\cdot\nu}, \nu \in Z^n \cap \lambda D\}.$$

Let q be a smooth function on T^n and let $M_q : L^2(T^n) \to L^2(T^n)$ be the linear mapping "multiplication by q."

THEOREM 2. *If μ_λ is the spectral measure of $\pi_\lambda M_q \pi_\lambda$, then, as λ tends to infinity, μ_λ/λ^n converges weakly to the measure, μ, defined by*

$$\mu(f) = \frac{\text{vol } D}{(2\pi)^n} \int_{T^n} f(q(x))\, dx .$$

Proof. Let

$$\lambda D = \{\xi \in \mathbf{R}^n, p(\xi) \leq \lambda\},$$

p a smooth homogeneous function of degree one on $\mathbf{R}^n - \{0\}$. Apply Theorem 1 to the operator $P: C^\infty(T^n) \to C^\infty(T^n)$ defined by $Pe^{i x\cdot\nu} = p(\nu) e^{i x\cdot\nu}$ for all $\nu \in Z^n$. Q.E.D.

This shows that the crude "space cut-off" described above is a reasonable way to renormalize the spectral measure of M_q. Except for the harmless factor, vol D, this renormalization is independent of the choice of D.

Widom's most recent work on the classical Szegö theorem has to do with the higher order asymptotics of the sequence trace $\pi_k M_q \pi_k$. We unfortunately don't have time here to touch on this subject, except to note that if P and B are as in Theorem 1 and $\phi(\lambda) = \text{trace}\ \pi_\lambda B \pi_\lambda$ then

$$\gamma(t) = \int_0^\infty e^{i\lambda t} \frac{d\phi}{d\lambda}(\lambda)\, d\lambda$$

admits an asymptotic expansion about $t = 0$ (of which Lemma 1 supplies the leading term). It is possible that this asymptotics is related in some devious way to Widom's asymptotics. (See [33] and [34]. See also [35].)

4. *Szegö-type formulas II*

Today I want to describe some complex variable analogues of the classical Szegö theorem. The results described below were obtained jointly by Boutet de Monvel and myself. Recently Weinstein [32] obtained some striking results on the spectral properties of the Schroedinger operator on S^n. Our results are Toeplitz operator analogues of Weinstein's results.

Let X be an n-dimensional non-singular complex projective variety embedded in $\mathbb{C}P^N$. Let L be the canonical line bundle on X. If we give $\mathbb{C}^{N+1}$ its standard Hermitian structure, $\langle z, w\rangle = \Sigma z_i \bar{w}_i$, then L acquires a Hermitian inner product. In addition there is a unique connection on L with the property that if s is a (local) non-vanishing homomorphic section

$$Ds = \bar{\partial} \ln |s|^2 \otimes s .$$

If ω is the curvature form of this connection, then (X, ω) is a Kaehler manifold. We will denote by ν the volume form $\omega \wedge \cdots \wedge \omega$ (n wedges).

Let X be defined by the system of equations:

$$(4.1) \qquad \begin{aligned} f_1(x_1, \cdots, x_{N+1}) &= 0 \\ &\vdots \\ f_r(x_1, \cdots, x_{N+1}) &= 0 \end{aligned}$$

where $f_1, \cdots, f_r$ are homogeneous functions in $\mathbb{C}[x_1, \cdots, x_{N+1}]$ and the ideal $(f_1, \cdots, f_r)$ is reduced. The k-th graded component of the graded ring, $\mathbb{C}[x_1, \cdots, x_{N+1}]/(f_1, \cdots, f_r)$ can be identified with the space of holomorphic sections of the line bundle $\overset{k}{\otimes} L^*$. For large k the dimension of this space is

$$(\text{degree } X) \frac{k^n}{n!} + O(k^{n-1}) . \tag{4.2}$$

(See [24].) We will refer back to this identity below.

The Hermitian inner product on L induces Hermitian inner products on all the $\overset{k}{\otimes} L^*$. Since X itself is equipped with the volume form ν the space of smooth sections of $\overset{k}{\otimes} L^*$ has an intrinsic pre-Hilbert structure. Let π_k be the orthogonal projection of this space onto the subspace of holomorphic sections.

THEOREM 1. *Given a smooth real-valued function,* q, *on* X *let* M_q *be the operator, "multiplication by* q,*" on the space of smooth sections of* $\overset{k}{\otimes} L^*$ *and let* μ_k *be the spectral measure of* $\pi_k M_q \pi_k$. *Then, as* k *tends to infinity,* $\frac{\mu_k}{k^n}$ *converges weakly to the measure,* μ, *defined by*

$$\mu(f) = \frac{1}{n!\gamma_n} \int_X f(q(x))\, d\nu, \qquad \forall f \in C(\mathbb{R}) . \tag{4.3}$$

where γ_n = volume $\mathbb{CP}^n$.

REMARK. If f is the constant function, 1, $\mu_k(f)$ is just the dimension of the space of holomorphic sections of $\overset{k}{\otimes} L^*$ and $\mu(f)$ is the volume of X times the factor $(n!\gamma_n)^{-1}$. In view of (4.2):

$$\text{degree } X = \text{volume } X/\text{volume } \mathbb{CP}^n$$

This formula is well known. (See [24].)

We will sketch a proof of Theorem 1, omitting a few technical details. As we remarked above, L is equipped with a Hermitian inner product.

Let W be the unit disk bundle in L, and ∂W the unit circle bundle. It is well known that W is a *strictly pseudo-convex domain.* This fact will play an essential role in our proof. To exploit it we will need a few general results about boundaries of pseudo-convex domains. For the moment let W be any compact strictly pseudo-convex domain with smooth boundary. Let $\dim W = n+1$. At each point $x \in \partial W$, $T_x \partial W \otimes \mathbb{C}$ contains a distinguished subspace spanned by the holomorphic vectors which are tangent to the boundary. The annihilator of this space in $T_x^* \partial W$ is one-dimensional and is called the space of *characteristic co-vectors* in T_x^*. We will denote by Σ the manifold of all characteristic co-vectors and by Σ^+ and Σ^- the connected components of $\Sigma - 0$ associated with the inner and outer orientations of the boundary. Σ^+ is a symplectic submanifold of $T^*\partial W$, and, of course, also a principal fiber bundle over ∂W with fiber R^+.

The L^2 closure in $L^2(\partial W)$ of the space of boundary values of holomorphic functions is the *Hardy space*, H^2, and the orthogonal projection of L^2 onto H^2 is the *Szegö projector*, S. An operator $T: H^2 \to H^2$ is a *Toeplitz* operator of order k if it can be written in the form $T = SQS$ where Q is a standard pseudo-differential operator of order k. It can be shown (with some effort) that Toeplitz operators form a ring under composition. The *symbol* $\sigma(T)$ of a Toeplitz operator T is the restriction to Σ^+ of $\sigma(Q)$ where $T = SQS$. These symbols have properties very similar to the symbols of pseudo-differential operators, viz:

i) $\sigma(T_1)\sigma(T_2) = \sigma(T_1 T_2)$

ii) $\sigma([T_1, T_2]) = \{\sigma(T_1), \sigma(T_2)\}$

iii) T of order k and $\sigma(T) = 0 \Longrightarrow$ T of order k-1.

From these symbolic properties one easily deduces:

LEMMA 1. *Let* T *be a* k-th *order Toeplitz operator which is elliptic, i.e.* $\sigma(T) \neq 0$ *everywhere; then there exists a* (-k)-th *order Toeplitz operator,* T′, *such that* $I - TT'$ *and* $I - T'T$ *are smoothing.*

LEMMA 2. *Let* T *be a self-adjoint, elliptic, Toeplitz operator of order* $k > 0$. *Then the spectrum of* T *is discrete and has no finite points of accumulation.*

Proof. Let T' be a parametrix of T. Since T' is of order $-k$, it is of the form $SQ'S$ where Q' is of order $-k < 0$, hence, compact. Thus T' is compact; hence its spectrum is discrete, bounded and has only zero as a point of accumulation. This implies that T has the stated properties.

Q.E.D.

Let's now go back to the line bundle $L \to X$ and the circle bundle, ∂W, in L. The action $\theta \to \phi_\theta$, of the circle group on ∂W is the restriction to ∂W of a holomorphic map of W onto itself, so $\phi_\theta^*: L^2 \to L^2$ preserves H^2 and defines on H^2 a one-parameter unitary group, $U(\theta)$, whose infinitesimal generator is a Toeplitz operator of order one, viz: $S \frac{1}{\sqrt{-1}} \frac{\partial}{\partial\theta} S$. We will denote this operator by $T^{\#}$.

LEMMA 3. $T^{\#}$ *is elliptic and self-adjoint. Its spectrum consists of the non-negative integers, the eigenspace corresponding to the eigenvalue* k *being identical with the space of holomorphic sections of* $\overset{k}{\otimes} L^*$.

Proof. At each point $x \in \partial W$ the vector $\frac{\partial}{\partial\theta}$ lies in a complement of the space spanned by the holomorphic and anti-holomorphic vectors so the symbol of $T^{\#} = S \frac{1}{\sqrt{-1}} \frac{\partial}{\partial\theta} S$ is everywhere non-zero. This shows that $T^{\#}$ is elliptic. $T^{\#}$ is obviously self-adjoint. By Lemma 2 the spectrum of $T^{\#}$ is discrete and has no finite points of accumulation. Since $\exp \sqrt{-1}\, 2\pi\, T^{\#} = I$ the spectrum must consist of integers. Let f be in the k-th eigenspace and let $x \in X$. Let D_x be the unit disk in L_x and let z be a holomorphic coordinate function on D_x. Then $f|\partial D_x$ lies in the Hardy space of ∂D_x and continues to satisfy $\frac{1}{\sqrt{-1}} \frac{d}{d\theta} f(\theta) = kf(\theta)$. This shows that k is non-negative and $f|\partial D_x$ is the restriction to ∂D_x of a function, az^k, where a depends smoothly on x as ∂D_x varies.

From the fact that f satisfies the boundary Cauchy-Riemann equations, it follows that a is holomorphic in x, so f is a holomorphic section of $\overset{k}{\otimes} L^*$ as claimed. Q.E.D.

Let q be a smooth function on X, $q_1 = \pi^* q$ where $\pi : \partial W \to X$ is projection, and $T_q = SM_{q_1}S$ where M_{q_1} is multiplication by q_1. Since M_{q_1} commutes with the circle group action, $U(\theta)$ and T_q commute. Let $f(T_q)$ be a polynomial function of T_q and consider the trace of $f(T_q)U(\theta)$. By Lemma 3

$$\text{trace } f(T_q)U(\theta) = \sum_{k=0}^{\infty} \mu_k(f)\, e^{ik\theta} \tag{4.4}$$

μ_k being the spectral measure of $\pi_k M_q \pi_k$. To obtain an analytic expression for this trace, we exploit the fact that $f(T_q)$ is a zeroth order Toeplitz operator with symbol $f(q)$. From functorial properties of "Hermite distributions" ([2] and [3]) one can show that

$$\text{trace } f(T_q)U(\theta) = \sum_{r=n}^{-N} c_r \chi_r(\theta) \quad \text{mod } C^{N-1}(S^1) \tag{4.5}$$

where $\chi_r(\theta)$ is the distribution

$$\sum_{k>0} k^r e^{ik\theta} . \tag{4.6}$$

The coefficient, c_n, of the leading term in (4.5) can be computed using, for instance, the symbol calculus for Hermite distributions developed in [4]. Up to a universal constant it is the integral of the symbol of $f(T_q)$ over the subset, $\sigma(T^{\#})(x,\zeta) = 1$, of Σ^+. However, the symbol of $f(T_q)$ is $f(q)$; so up to universal constant this is just the integral on the right hand side of (4.3). Comparing the k-th coefficients of (4.5) and (4.4), we get (4.3). Q.E.D.

The details of this proof will appear elsewhere. I'll devote the second part of this lecture to discussing a corollary of Theorem 1 having to do with the asymptotic behavior of representations of compact Lie groups. Let G be a compact semi-simple Lie group and H a maximal torus in G. Let $\mathfrak{g}$ and $\mathfrak{h}$ be the associated Lie algebras. Given an irreducible representation $\rho: G \to U(N)$ the restriction of ρ to H breaks up into irreducible one-dimensional representations, $\chi: H \to U(1)$. The infinitesimal representations $d\chi: \mathfrak{h} \to \sqrt{-1}\,\mathbf{R}$ are of the form $d\chi = \sqrt{-1}\,\beta$ is in the (real) dual space of $\mathfrak{h}$. These β's are called the *weights* of the representation, ρ. Let $\mathfrak{b}^+$ be a fixed Borel subalgebra of $\mathfrak{g} \otimes \mathbf{C}$. A weight, β, is called a *dominant weight* of the representation ρ if the subspace corresponding to it is $\mathfrak{b}^+$-invariant. Every irreducible representation possesses a unique dominant weight, and it always occurs with multiplicity one. The representation is completely determined by its dominant weight. Moreover there exists a basis $\beta_1, \cdots, \beta_r$ of $\mathfrak{h}^*$ such that $\beta \in \mathfrak{h}^*$ is the dominant weight of an irreducible representation if and only if $\beta = \Sigma\, m_i \beta_i$ where the m_i's are non-negative integers.

We will fix some $\beta = \Sigma\, m_i \beta_i$. For each $k > 0$, there is a unique irreducible representation, V_k, with $k\beta$ as its dominant weight. The theorem we are about to state describes the asymptotic distribution of the weights of V_k as k gets large. Before we state this theorem, we will need another description of the irreducible representations of G in terms of the so-called "orbit picture": G acts on $\mathfrak{g}^*$ by the dual of the adjoint representation. It was observed by Kostant that the tri-linear pairing

$$\mathfrak{g} \times \mathfrak{g} \times \mathfrak{g}^* \to \mathbf{R}, \quad (x, y, \alpha) \to \alpha[x, y]$$

induces a symplectic structure on each orbit of G in $\mathfrak{g}^*$. More generally let X be any symplectic manifold. The C^∞ functions on X form a Lie algebra, P, under Poisson bracket. If $\mathfrak{H}$ is the Lie algebra of symplectic vector fields on X there is a short exact sequence

$$0 \longrightarrow \mathbf{C} \longrightarrow P \xrightarrow{\gamma} \mathfrak{H} .$$

where $\gamma(f)$ is the Hamiltonian vector field associated with f. If the Lie group, G, acts on X as a group of symplectic diffeomorphisms we get a map $\mathfrak{g} \to \mathfrak{H}$. If G is semi-simple there is a unique lifting:

$$\begin{array}{ccc} & & P \\ & \overset{\rho}{\nearrow} & \big\downarrow \gamma \\ \mathfrak{g} & \longrightarrow & \mathfrak{H} \end{array}$$

By duality one gets a map

$$\rho^t : X \to \mathfrak{g}^* \tag{4.7}$$

where $\rho^t(x) \in \mathfrak{g}^*$ evaluated on $v \in \mathfrak{g}$ is $\rho(v)(x)$. The map (4.7) is called the "moment map" by Kostant and Souriau. Its image is a union of orbits in $\mathfrak{g}^*$, and, if G acts transitively on X; its image is a single orbit; and it is locally a symplectic diffeomorphism.

In particular, let G be compact and consider an irreducible representation of G on a vector space V, with maximal weight β. G also acts on the projective space, $V - 0/\mathbb{C}^*$. The one-dimensional subspace of V corresponding to β determines a point in this projective space, and the orbit, X, of G through this point turns out to be a projective subvariety. Since its Kaehler structure is invariant under the action of G, X is a transitive symplectic G-space, and its moment map determines an orbit of G in $\mathfrak{g}^*$. This sets up a 1-1 correspondence between irreducible representations of G and *integral* orbits of G in $\mathfrak{g}^*$, (an orbit being integral if its symplectic two-form represents an integral cohomology class).

One last definition: A representation is *generic* if the corresponding orbit in $\mathfrak{g}^*$ is of codimension equal to the dimension of H.

THEOREM 2. *Let O be the orbit corresponding to a generic irreducible representation with dominant weight, β. Let μ be the direct image of the symplectic measure on O under the mapping*

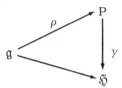

$$0 \longrightarrow \mathfrak{g}^* \xrightarrow{i^*} \mathfrak{h}^*$$

where $i : \mathfrak{h} \to \mathfrak{g}$ *is the inclusion map. Let* μ_k *be the discrete measure*

$$\sum \delta\left(x - \frac{\beta_i}{k}\right)$$

on $\mathfrak{h}^*$, *the sum taken over all the weights of the representation of* G *on* V_k. *Finally let* n *be the dimension of* G/H. *Then* μ_k/k^n *converges weakly to a constant multiple of* μ *as* k *tends to infinity, the constant depending only on* n.

Proof. Let X be the complex projective variety associated with the representation $V = V_1$ (see supra). By Borel-Weil, V_k can be identified with the space of holomorphic sections of $\overset{k}{\otimes} L^*$. As in the proof of Theorem 1, let W be the disk bundle of L and ∂W the circle-bundle. G acts on ∂W and this action preserves the Hardy space and commutes with the circle group action. Given $v \in \mathfrak{g}$ let $T_v = SD_vS$ where S is the Szegö projector, and D_v the first order differential operator associated with the infinitesimal action of v. The map, $v \to T_v$, is a representation of $\mathfrak{g}$ in the Lie algebra of first order Toeplitz operators on ∂W. As in the proof of Theorem 1 let $T^{\#}$ be the Toeplitz operator, $S \frac{1}{\sqrt{-1}} \frac{\partial}{\partial \theta} S$. Then

$$\sigma(T_v) = \rho(v)\sigma(T^{\#}) \tag{4.8}$$

for a smooth function, $\rho(v)$, on ∂W. Since both sides of (4.8) are invariant under the circle group $\rho(v)$ is actually a function on X, and the map, $v \to \rho(v)$, is a homomorphism of $\mathfrak{g}$ into the Poisson algebra of X. It is easy to check that this homomorphism lifts the canonical homomorphism of $\mathfrak{g}$ into the symplectic algebra of X; so by the theorem of Kostant quoted above:

LEMMA 4. *The map* $x \to \rho^t(x)$, *is the moment map and its image is the orbit in* $\mathfrak{g}^*$ *associated with* V.

Now let $v_1, \cdots, v_r$ *be a basis of* $\mathfrak{h}$. *Set* $T_i = (T_{v_i})(T^{\#})^{-1}$. *The* T_i*'s commute and* $\sigma(T_i) = q_i = \rho(v_i) \in C^\infty(X)$. *For every polynomial function,* f, *in the variables* $x_1, \cdots, x_r$ *set* $f(T) = f(T_1, \cdots, T_r)$. *This is a zeroth order Toeplitz operator with symbol* $f(q_1, \cdots, q_r)$.

LEMMA 5. *Let* π_k *be the orthogonal projection of the Hardy space* $H^2(\partial W)$ *onto the* k-th *eigenspace of* $T^{\#}$. *Then*

$$\lim \operatorname{trace} \pi_k f(T) \pi_k = \frac{1}{\gamma_n n!} \int_K f(q_1, \cdots, q_n)\, d\nu \tag{4.9}$$

Proof. Let $\overline{T}_i = SM_{q_i}S$ where M_{q_i} is "multiplication by q_i." Then T_i differs from $\overline{T}_i$ by a compact operator; so $f(T)$ differs from $f(\overline{T})$ by a compact operator. Since rank $\pi_k = O(k^n)$ (see 4.2) the expression

$$\frac{1}{k^n} \operatorname{trace} \pi_k(f(T) - f(\overline{T}))\pi_k$$

tends to zero. However, by Theorem 1, $k^{-n} \operatorname{trace} \pi_k f(\overline{T}) \pi_k$ tends to the right hand side of (4.9) as k tends to infinity. Q.E.D.

By Borel-Weil the k-th eigenspace of $T^{\#}$ is the space, V_k, and on this space

$$\operatorname{trace} f(T) = \operatorname{trace} f(T_{v_1}(T^{\#})^{-1}, \cdots, T_{v_r}(T^{\#})^{-1}) .$$

The second term is just

$$\sum f\left(\frac{\tilde{\beta}_i}{k}\right)$$

where β_i, $i = 1, \cdots, \dim V_k$, are the weights of V_k and $\tilde{\beta}_i = (\beta_i(v_1), \cdots, \beta_i(v_r))$. But this is just $\mu_k(f)$ where μ_k is the measure

$$\sum \delta\left(x - \frac{\beta_i}{k}\right).$$

By Lemma 4, the right hand side of (4.9) is the direct image of the symplectic measure on X under the moment map. Comparing the two sides of (4.9) we get Theorem 2. Q.E.D.

It is well known that if an irreducible representation of G has dominant weight, β, all the weights of the representation lie in the convex hull of the orbit of β under the action of the Weyl group. On the other hand, Kostant has proved the following result.

THEOREM 3. *If O is an orbit in $\mathfrak{g}^*$ corresponding to a non-degenerate irreducible representation with maximal weight β, the image of O in $\mathfrak{h}^*$ is equal to the convex hull of the orbits of β under the action of the Weyl group.*

Let $\beta_1, \cdots, \beta_{n_k}$ be the weights of V_k. According to Theorem 2 and Theorem 3, as k goes to infinity, the points

$$\frac{\beta_1}{k}, \cdots, \frac{\beta_{n_k}}{k}$$

become dense in the convex hull of the orbit of β under the action of the Weyl group. Figure 1 below is the image of a generic orbit of SU(3) in $\mathfrak{h}^* = R^2$.

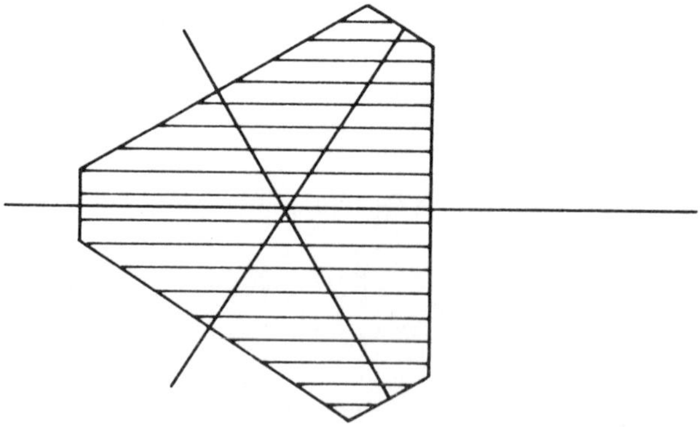

Fig. 1

REFERENCES

[1] Agmon, S., On kernels, eigenvalues and eigenfunctions of operators related to elliptic problems. Comm. Pure and Appl. Math. 18 (1965), 627-663.

[2] Avakumovic, V. G., Uber die Eigenfunktionen auf geschlossenen Riemannscnen Mannigfaltigkeiten. Math. Z. 65 (1956), 327-344.

[3] Bérard, P., On the wave equation on a compact Riemannian manifold without conjugate points. Math. Z. 155 (1977), 249-276.

[4] Borel, A., and Weil, A., Représentations lineaires et espaces homogènes Kahleriens des groupes de Lie compacts. Séminaire Bourbaki, May 1954.

[5] Boutet de Monvel, L., Hypoelliptic operators with double characteristics and related pseudodifferential operators. Comm. Pure Appl. Math. 27 (1974), 585-639.

[6] Brüning, J., Zur Abschätzung der Spectralfunction elliptischer Operatoren. Math. Z. 137 (1974), 74-85.

[7] Chazarain, J., Formule de Poisson pour les varietes Riemanniennes. Invent. Math. 24 (1974), 65-82.

[8] Colin de Verdière, Y., Nombre de points entiers dans une famille homothetique de domaines de $\mathbf{R}^n$, (to appear).

[9] Courant, R., Über die Eigenwerte bei den Differentialgleichungen der mathematischen Physik. Math. Z. 7 (1920), 1-57.

[10] Duistermaat, J. J., and Guillemin, V., The spectrum of positive elliptic operators and periodic geodesics. Invent. Math. 29 (1975), 39-79.

[11] Guillemin, V., Symplectic spinors and partial differential equations. Proceedings, C.N.R.S. Colloque sur la Geometrie Symplectique, Aix en Provence, June 1974.

[12] ———, A symbol calculus for Hermite operators. MIT preprint.

[13] Helton, W., An operator algebra approach to partial differential equations, propagation of singularities and spectral theory. Indiana J. Math. 26 (1977), 997-1018.

[14] Herz, C., On the number of lattice points in a convex set. Amer. J. Math. 84 (1962), 126-133.

[15] Hlawka, E., Über Integrale auf konvexen Körpern. Monats h. f. Math. 54 (1950), 1-36.

[16] Hörmander, L., The spectral function of an elliptic operator. Acta Math. 121 (1968), 193-218.

[17] Kac, M., Toeplitz matrices, translation kernels, and a related problem in probability theory. Duke Math. J. 21 (1954), 501-509.

[18] Kostant, B., Quantization and unitary representations. Lecture Notes in Math. 170 Springer Verlag (1970), 87-208.

[19] Landau, E., Vorlesungen über Zahlentheorie. Chelsea, New York (1969).

[20] Levitan, B. M., On the eigenfunction expansion of a self-adjoint equation with partial derivatives. Trudy Moskov. Math. Obshch. 5(1956), 269-298.

[21] ———, On the eigenfunction expansion of the equation $\Delta u + \{\lambda - q(x_1, \cdots, x_n)\}u = 0$. Izv. Akad. Nauk. SSSR Ser. Mat. 20(1956), 437-468.

[22] ———, Asymptotic behavior of the spectral function of an elliptic equation. Russian Math. Surveys 26, no. 6(1971), 165-232.

[23] ———, On a special Tauberian theorem. Izv. Akad. Nauk. SSSR Ser. Mat. 17(1953), 269-284.

[24] Mumford, D., Algebraic Geometry I: Complex Projective Varieties. Springer Verlag, New York (1976).

[25] Petrowsky, I. G., Lectures on Partial Differential Equations. Interscience, New York, 1954.

[26] Randol, B., A lattice point problem I. Trans. AMS 121(1966), 257-268.

[27] ———, A lattice point problem II. Trans. AMS 125(1966), 101-113.

[28] Seeley, R., A sharp asymptotic remainder estimate for the spectrum of the Laplacian in a domain in $\mathbf{R}^3$, (to appear).

[29] Szegö, G., Beiträge zur Theorie der Toeplitzschen Formen. Math. Z. 6(1920), 167-202.

[30] ———, On certain Hermitian forms associated with the Fourier series of a positive function. Comm. Sém. Math. Univ. Lund, tome suppl. (1952), 228-237.

[31] van der Corput, J. G., Zahlentheoretische Abschätzungen mit Andwendengen auf Gitterpunkt probleme. Math. Z. 17(1923), 250-259.

[32] Weinstein, A., Asymptotics of eigenvalue clusters for the Laplacian plus a potential, (to appear).

[33] Widom, H., Asymptotic behavior of block Toeplitz matrices and determinants. Advances in Math. 13(1974), 284-322.

[34] ———, Asymptotic behavior of block Toeplitz matrices and determinants II. Advances in Math. 21(1976), 1-29.

[35] ———, Asymptotic expansions of determinants for families of trace class operators, (to appear).

[36] ———, Families of pseudodifferential operators. Advances in Math., (to appear).

[37] Widom, H., Eigenvalue distribution theorems in certain homogeneous spaces, (to appear).

[38] Weyl, H., Über die asymptotische Verteilung der Eigenwerte. Göttinger Nachr. (1911), 110-117.

SZEGÖ'S THEOREM AND A COMPLETE SYMBOLIC CALCULUS FOR PSEUDO-DIFFERENTIAL OPERATORS

Harold Widom

1. *Introduction*

The classical Szegö theorem [10; 3, §5.2] states that for any bounded real-valued function ϕ on the unit circle, with Fourier coefficients denoted $\hat{\phi}_j$, the eigenvalues of the Toeplitz matrix

$$T_N = (\hat{\phi}_{i-j}) \quad 0 \leq i,\ j \leq N$$

are asymptotically distributed as the values of ϕ in the sense that for any continuous function F

$$\lim_{N \to \infty} N^{-1} \sum_{\lambda \in \text{spec } T_N} F(\lambda) = \frac{1}{2\pi} \int_0^{2\pi} F[\phi(e^{i\theta})]\, d\theta\ .$$

Szegö proved the general result from the special case $F(\lambda) = \log \lambda$ (assuming of course that $\phi > 0$), the assertion then being equivalent to

$$\log \det T_N = N\, \widehat{\log \phi}(0) + o(N)\ .$$

This has since been generalized in many directions. Szegö himself [11; 3, §5.5] showed that if ϕ is sufficiently smooth then the error term o(N) is in fact

Seminar on Singularities of Solutions
0-691-08221-9/78/0261 $01.15/1 (cloth)
0-691-08213-8/78/0261 $01.15/1 (paperback)
For copying information, see copyright page

$$\sum_{k=1}^{\infty} k \widehat{\log \phi}(k) \widehat{\log \phi}(-k) + o(1) .$$

M. Kac [6] proved the continuous analogue of this, whereby the matrix T_N is replaced by an operator of the form

$$A_\alpha : f(x) \to f(x) + \int_0^\alpha k(x-y) f(y) \, dy$$

on $L_2(0,\alpha)$, and α is a large parameter. Kac showed that if one defines

$$\sigma(\xi) = \int e^{-i\xi z} k(z) \, dz$$

$$s(z) = \frac{1}{2\pi} \int e^{i\xi z} \log(1+\sigma(\xi)) \, d\xi$$

then under suitable conditions

$$\log \det A_\alpha = \alpha s(0) + \int_0^\infty z s(z) s(-z) \, dz + o(1)$$

as $\alpha \to \infty$. There is a higher-dimensional analogue of this. If

$$A_\alpha : f(x) \to f(x) + \int_{\alpha\Omega} k(x-y) f(y) \, dy \tag{1.1}$$

on $L_2(\alpha\Omega)$, with Ω a smooth bounded subset of R^n, then

$$\log \det A_\alpha = a_0 \alpha^n + a_1 \alpha^{n-1} + o(\alpha^{n-1})$$

with

$$a_0 = s(0) \operatorname{vol} \Omega$$

$$a_1 = \frac{1}{4} \int_{\partial\Omega} \int_{R^n} |z \cdot n| s(z) s(-z) \, dz dx .$$

Here $s(z)$ is defined in anology with the one-dimensional case, dx denotes surface measure on $\partial\Omega$, and n is the unit inner normal to Ω at x.

This lecture describes how, via a complete symbolic calculus for pseudo-differential operators on manifolds, one can extend this to an asymptotic expansion

$$\log \det A_\alpha \sim a_0\alpha^n + a_1\alpha^{n-1} + a_2\alpha^{n-2} + \cdots \tag{1.2}$$

and, in principle, compute as many of the coefficients a_k as desired. For the curious, we state the result

$$a_2 = -\frac{1}{4}\int_{\partial\Omega}\int_{R^n} |z\cdot n|\, L(z_t \otimes z_t)\, \frac{\partial s(z)}{\partial n}\, s(-z)\, dz dx$$

$$+\frac{1}{2}\int_{\partial\Omega}\int_{R^n} z\cdot n\, |z\cdot n|\, L(z_t \otimes \operatorname{grad}_t s(z))\, s(-z)\, dz dx$$

where z_t denotes the component of z tangent to $\partial\Omega$ at x, $\partial s(z)/\partial n$ denotes the derivative in the direction of the inner normal, and L is the second fundamental form of $\partial\Omega$ with respect to the inner normal.

The symbolic calculus has other by-products of which we mention two. (1) Another derivation of the heat expansion for the Laplace-Beltrami operator on a compact Riemannian manifold, in which it is manifest that the coefficients are polynomials in the curvature tensor and its covariant derivatives.[1)] (2) A functional calculus for pseudo-differential operators where it is shown, for certain functions f and pseudo-differential operators A, that $f(A)$ is also pseudo-differential; and the complete symbol of $f(A)$ is determined in terms of that of A.

Very few details and no complete proofs will be given here. For these we refer to [13; 14; 15]. Rather we describe the main ideas that go into

[1)] This fact is implicitly contained in [7] and even in the earlier work by Hadamard [4].

the development leading eventually to the expansion (1.2) and similar expansions. In §2 we consider a simpler problem which is solved by consideration of certain families of pseudo-differential operators in R^n. In §3 an analogous half-space problem is solved by adding a modification of the classical Wiener-Hopf factorization. The next section contains the symbolic calculus on manifolds. The expansion (1.2) is gotten by using this calculus, thinking of Ω as a manifold with boundary and applying the technique of §2. The final sections contain brief discussions of the two applications of the symbolic calculus mentioned above.

2. *Pseudo-differential families in* R^n

Suppose $\sigma(x,\xi)$ is a symbol of order $\omega < -n$ with compact x-support. Then the pseudo-differential operator with symbol σ,

$$f(x) \to \int \hat{f}(\xi)\, e^{i\xi\cdot x} \sigma(x,\xi)\, d\xi \ ,$$

is of trace class with trace

$$(2\pi)^{-n} \iint \sigma(x,\xi)\, d\xi\, dx \ .$$

In fact if

$$k(x,z) = (2\pi)^{-n} \int e^{i\xi\cdot z} \sigma(x,\xi)\, d\xi$$

then the operator has kernel $k(x, x-y)$ so its trace equals

$$\int k(x,0)\, dx = (2\pi)^{-n} \iint \sigma(x,\xi)\, d\xi\, dx \ .$$

Similarly the operator

$$T_a : f(x) \to \int \hat{f}(\xi)\, e^{i\xi\cdot x} \sigma(x,\xi/a)\, d\xi \tag{2.1}$$

has kernel $a^n k(x, a(x-y))$ and trace

$$(2.2)\qquad \left(\frac{\alpha}{2\pi}\right)^n \iint \sigma(x,\xi)\, d\xi\, dx\ .$$

Note that the operator A_α of (1.1) is, by an obvious change of variable, unitarily equivalent to $P(I+T_\alpha)P$ if P is multiplication by the characteristic function of Ω and σ is independent of x in a neighborhood of Ω. This P complicates matters. Without it the expansion for

$$\log \det (I+T_\alpha) = \operatorname{tr} \log (I+T_\alpha)\ ,$$

or more generally for $\operatorname{tr} F(T_\alpha)$ for certain analytic functions F, is a simple consequence of the calculus of pseudo-differential operators on R^n, by a method we now describe.

The well-known formula for the symbol of a product is

$$(2.3)\qquad \sigma_{AB} = \sum \frac{i^{-k}}{k!} \frac{\partial^k \sigma_A}{\partial \xi^k} \frac{\partial^k \sigma_B}{\partial x^k}\ ,$$

the sum being taken over all multi-indices k.[1] The symbol of T^{-1}, the inverse of T modulo smoothing operators, is obtained from this as follows. Define

$$P_m(\sigma,\tau) = \sum_{|k|=m} \frac{i^{-k}}{k!} \frac{\partial^k \sigma}{\partial \xi^k} \frac{\partial^k \tau}{\partial x^k}$$

and, inductively,

$$(2.4)\qquad Q_0(\sigma) = \sigma^{-1},\ \ Q_m(\sigma) = -\sigma^{-1} \sum_{r=1}^{m} P_r(\sigma, Q_{m-r}(\sigma))\ .$$

Under a suitable ellipticity condition, if $\sigma_T = \sigma$, then

$$\sigma_{T^{-1}} = \sum_{m=0}^{\infty} Q_m(\sigma)\ .$$

---

[1] The interpretation of (2.3) is that the difference between σ_{AB} and the sum of sufficiently many terms of the series is a symbol of arbitrarily low negative order. Similar identities to follow should be interpreted in the same way.

For each $m > 0$ one can write

$$Q_m(\sigma) = \sum_{k=2}^{2m} (-1)^k \sigma^{-k-1} Q_{m,k}(\sigma) \tag{2.5}$$

where the $Q_{m,k}$ are certain differential operators satisfying

$$Q_{m,k}(\sigma-\lambda) = Q_{m,k}(\sigma)$$

for each λ. Thus

$$\sigma_{(T-\lambda)^{-1}} = (\sigma-\lambda)^{-1} + \sum_{m=1}^{\infty} \sum_{k=2}^{2m} (-1)^k (\sigma-\lambda)^{-k-1} Q_{m,k}(\sigma) .$$

If F is analytic then multiplication by $-F(\lambda)/2\pi i$ and integration over a suitable contour give

$$\sigma_{F(T)} = F(\sigma) + \sum_{m=1}^{\infty} \sum_{k=2}^{2m} \frac{1}{k!} F^{(k)}(\sigma) Q_{m,k}(\sigma) . \tag{2.6}$$

All this is of course purely formal but may be justified in many cases. Let us see what happens if we apply this to families such as (2.1). If we agree to call $\sigma(x,\xi)$ the symbol of the family then the analogue of (2.3) is

$$\sigma_{AB} = \sum_{m=0}^{\infty} a^{-m} P_m(\sigma,\tau) .$$

Following the preceding discussion through shows that the analogue of (2.6) for families is

$$\sigma_{F(T_a)} = F(\sigma) + \sum_{m=1}^{\infty} a^{-m} \sum_{k=2}^{2m} \frac{1}{k!} F^{(k)}(\sigma) Q_{m,k}(\sigma) . \tag{2.7}$$

And combining this with (2.2) gives the asymptotic expansion for $\operatorname{tr} F(T_a)$.

Since

$$Q_{1,2}(\sigma) = -i \sum_{|k|=1} \frac{\partial^k\sigma}{\partial\xi^k}\frac{\partial^k\sigma}{\partial x^k} = -i\ \mathrm{grad}_\xi\sigma \cdot \mathrm{grad}_x\sigma$$

we find for the first two terms

$$\text{(2.8)}\quad \mathrm{tr}\ F(T_\alpha) = \left(\frac{\alpha}{2\pi}\right)^n \iint \left[F(\sigma) - \frac{i}{2}\alpha^{-1}F''(\sigma)\ \mathrm{grad}_\xi\sigma\cdot\mathrm{grad}_x\sigma + \cdots\right] d\xi\, dx\,.$$

In particular, for $F(\lambda) = \log(1+\lambda)$, we find that

$$\log\det(I+T_\alpha) =$$

$$\alpha^n \int s(x,0)\,dx + \frac{\alpha^{n-1}}{2}\iint s(x,z)\ z\cdot\mathrm{grad}_x\ s(x,-z)\,dz dx + O(\alpha^{n-2})$$

where

$$\text{(2.9)}\qquad s(x,z) = (2\pi)^{-n}\int e^{i\xi\cdot z}\log(1+\sigma(x,\xi))\,d\xi\ .$$

For this to hold, it suffices that there be a continuously defined

$$\log(1+\sigma(x,\xi))$$

which vanishes as $|\xi| \to \infty$. For it may be shown that then $\log(I+T_\alpha)$ is a pseudo-differential family with symbol given by the series (2.7) with $F(\lambda) = \log(1+\lambda)$.

3. *The half-space problem*

The problem is to find the asymptotic expansion, under suitable conditions, of

$$\log\det P(I+T_\alpha)P\ ,$$

where T_α is given by (2.1) and P is multiplication by the characteristic function of the half-space $x_n \geq 0$.

To give the main idea we consider first the case $n = 1$, assume σ is independent of x, and ignore the parameter α. Thus we look at

$$Tf(x) = \int_{-\infty}^{\infty} k(x-y)\,f(y)\,dy$$

with the view of evaluating $\det P(I+T)P$, where P is multiplication by the characteristic function of R^+. Actually, of course, this T is never of trace class and the determinant is undefined but we shall get to a point where, if it were defined, we could evaluate it.

We start with the easily checked fact

$$\text{support } k \subset R^+, \quad \text{support } f \subset R^+ \Longrightarrow \text{support } Tf \subset R^+$$

and similarly with R^+ replaced by R^- everywhere. Otherwise said,

$$\text{support } k \subset R^+ \Longrightarrow (I-P)TP = 0$$

$$\text{support } k \subset R^- \Longrightarrow PT(I-P) = 0\,.$$

Thus given T, convolution by k, if we define

$$T_+ = \text{convolution by } Pk$$

$$T_- = \text{convolution by } (I-P)k$$

then

(3.1) $$T = T_- + T_+, \quad (I-P)T_+P = 0, \quad PT_-(I-P) = 0\,.$$

For convolution operators, pseudo-differential operators with symbol independent of x, the symbol of a product is the product of the symbols exactly. It follows that if $\sigma(\xi)$, the symbol of T, has the property that there is a continuously defined

$$\log(1+\sigma(\xi))$$

vanishing when $\xi \to \pm\infty$, then

$$I + T = e^S$$

where S is the convolution operator with symbol $\log(1+\sigma)$. Now apply the decomposition (3.1) to S rather than T. Since S_- and S_+ are convolution operators they commute and so

$$I + T = e^{S_- + S_+} = e^{S_-} e^{S_+} .$$

This is the Wiener-Hopf factorization. Its usefulness derives from the fact that from

$$(I - P) S_+ P = 0$$

one deduces

$$e^{S_+} P = e^{PS_+P} ,$$

where in the exponential on the right one thinks of the projection P as the identity operator; and similarly

$$P e^{S_-} = e^{PS_-P} .$$

Hence

$$P(I+T)P = e^{PS_-P} e^{PS_+P} . \tag{3.2}$$

The usual use of the factorization is to invert $P(I+T)P$, the inverse being clearly

$$e^{-PS_+P} e^{-PS_-P} .$$

If T were trace class then so would be S and the multiplicativity of the determinant together with the general fact

$$\det e^A = e^{\operatorname{tr} A}$$

would give, by (3.2),

$$\log \det P(I+T)P = \operatorname{tr} PS_-P + \operatorname{tr} PS_+P = \operatorname{tr} PSP .$$

Now let us see what happens if T is the operator family T_α of (2.1) and P is multiplication by the characteristic function of the half-space. The kernel of T is

$$\alpha^n k(x, \alpha(x-y)) .$$

We define $T_{\pm}$ to be the operators with kernels

$$\alpha^n k_{\pm}(x, \alpha(x-y))$$

where for each fixed x

$$k_+(x,\cdot) = Pk(x,\cdot)$$
$$k_-(x,\cdot) = (I-P)\,k(x,\cdot)\ .$$

Then (3.1) still holds.

Next, assume there is a continuously defined

$$\log(1+\sigma(x,\xi))$$

vanishing when $|\xi| \to \infty$. By the remark at the end of §2 we can write

$$I + T = e^S$$

with $S = S_\alpha$ a pseudo-differential family whose complete symbol we can compute.

The first two steps therefore generalize, and we can apply the decomposition (3.1) to write

$$I + T = e^{S_- + S_+}\ .$$

But now there is a difficulty since S_- and S_+ do not commute and so

$$e^{S_- + S_+} \neq e^{S_-} e^{S_+}\ .$$

At this point therefore we apply the Baker-Campbell-Hausdorff formula [5, Chap. X] which tells us that, at least formally,

(3.3) $$e^A e^B = \exp\left\{A + B + \sum_{k=2}^{\infty} u_k(A,B)\right\}$$

where each $u_k(A,B)$ is a linear combination of $k-1$-fold commutators

$$[\cdots[X_1 X_2] X_3 \cdots X_k]$$

with each X_i equal to either A or B. In case A and B are pseudo-differential families of type (2.1)

$$\sum_{k=2}^{\infty} u_k(A, B)$$

turns out to be an asymptotic expansion with respect to α.

Now we look for a family $X = X_\alpha$ such that

$$e^{X_-} e^{X_+} = e^S .$$

For then we would have

$$P(I+T)P = e^{PX_-P} e^{PX_+P} ,$$

(3.4) $$\log \det P(I+T)P = \operatorname{tr} PXP = \left(\frac{\alpha}{2\pi}\right)^n \int_{R^n} \int_{x_n \geq 0} \sigma_X(x, \xi)\, dx d\xi .$$

To find X we use (3.3) and see that what we want is

$$S = X_- + X_+ + \sum_{k=2}^{\infty} u_k(X_-, X_+) = X + \sum_{k=2}^{\infty} u_k(X_-, X_+) .$$

Equivalently X is to be fixed under the mapping

$$X \to S - \sum_{k=2}^{\infty} u_k(X_-, X_+) ,$$

and the standard iteration procedure produces an asymptotic expansion for X,

(3.5) $$X = S - \frac{1}{2} [S_-, S_+] + \cdots ,$$

where the dots represent terms involving two or more commutations. Application of (3.4) then gives the asymptotic expansion

$$\log \det P(I+T_\alpha)P = a_0\alpha^n + a_1\alpha^{n-1} + a_2\alpha^{n-2} + \cdots .$$

The first two coefficients are easily found to be given by

$$a_0 = \int_\Omega s(x,0)\,dx$$

(3.6) $$a_1 = \frac{1}{2}\int_\Omega \int_{R^n} s(x,z)\, z\cdot \mathrm{grad}_x\, s(x,-z)\,dzdx + \frac{1}{4}\int_{\partial\Omega}\int_{R^n} |z_n|\, s(x,z)\, s(x,-z)\,dzdx$$

where s is given by (2.9), Ω is the half-space $x_n \geq 0$ and $\partial\Omega$ the hyperplane $x_n = 0$. The interior terms $\int_\Omega$ arise from the symbol of S, the boundary term $\int_{\partial\Omega}$ from that of $[S_-, S_+]$. Further terms can be found by straightforward laborious computation.

4. *Pseudo-differential operators on manifolds*

The symbol of a pseudo-differential operator A on R^n is given by

$$\sigma_A(x,\xi) = e^{-i\xi\cdot x} A\, e^{i\xi\cdot x} .$$

Equivalently

$$\sigma_A(x_0,\xi) = A\, e^{i\xi\cdot(x-x_0)}\Big|_{x=x_0} .$$

Applying $\partial^k/\partial\xi^k$ to both sides, multiplying by

$$\frac{i^{-k}}{k!} \frac{\partial^k f}{\partial x^k}\bigg|_{x=x_0}$$

and adding give, by Taylor's theorem,

$$\sum \frac{i^{-k}}{k!} \frac{\partial^k \sigma_A}{\partial \xi^k} \frac{\partial^k f}{\partial x^k}\bigg|_{x=x_0} = Af(x)\, e^{i\xi\cdot(x-x_0)}\bigg|_{x=x_0} .$$

Equivalently,

$$e^{-i\xi\cdot x} Af(x)\, e^{i\xi\cdot x} = \sum \frac{i^{-k}}{k!} \frac{\partial^k \sigma_A}{\partial \xi^k} \frac{\partial^k f}{\partial x^k} . \tag{4.1}$$

This is the basic asymptotic expansion from which, for example, (2.3) follows. The ingredients were the phase function $\xi\cdot(x-x_0)$ which is linear in both ξ and x, and Taylor's theorem which expands any function in powers of $x - x_0$ ($\partial/\partial\xi$ of the phase function) with coefficients involving the derivatives of f at x_0. It turns out that for any manifold M with a connection there is a natural substitute for these things. This leads to a complete symbolic calculus for operators on manifolds with formulas similar to those derived in §2.

A connection (covariant derivative) ∇ on M is, loosely speaking, a mapping taking tensor fields of type (r, s) to tensor fields of type $(r, s+1)$ such that

i) $\nabla f = df$ for a scalar function f;

ii) the product rule holds;

iii) ∇ commutes with contractions.

Our phase function will be a real-valued C^∞ function $\ell(v, x)$ ("ℓ" is for "linear") on $T^*M \times M$ satisfying

(a) for each x, $\ell(\cdot, x)$ is linear on fibers of T^*M;

(b) for each v, $\ell(v, \cdot)$ satisfies

$$\partial^k \ell(v, x)\big|_{x=\pi(v)} = \begin{cases} v, & k = 1 \\ 0, & k \neq 1 . \end{cases}$$

Here ∂^k denotes the symmetrization of the k-th covariant derivative $\nabla^k \ell$ taken with respect to x. Such functions exist, are clearly not unique, and need never be known explicitly. The interested reader may refer to [15] for details of this and what follows.

For fixed x and x_0 think of $\ell(\cdot, x)$ as a function on the fiber $T^*_{x_0}$. By (a) it is a linear function and so may be thought of as an element of T_{x_0}. We denote this element by $\ell(x_0, x)$; it is the analogue of the function $x - x_0$ on R^n. The k-fold tensor product

$$\ell(x_0, x)^k \overset{\text{def}}{=} \otimes_k \ell(x_0, x)$$

belongs to $\otimes_k T_{x_0}$ and so acts on any element of $\otimes_k T^*_{x_0}$ to produce a scalar. Since

$$\nabla^k f(x_0) \in \otimes_k T^*_{x_0}$$

we see that

$$\nabla^k f(x_0) \ell(x_0, x)^k$$

is, for each x, a scalar. In other words this is a scalar-valued function of x. The manifold version of Taylor's theorem [15, §2] states that

$$f(x) = \sum_{k=0}^{\infty} \frac{1}{k!} \nabla^k f(x_0) \ell(x_0, x)^k \tag{4.2}$$

in the sense that for each N

$$f - \sum_{k=0}^{N}$$

vanishes to order N + 1 at x_0.

The symbol of A, with respect to the connection ∇, is obtained as follows. We first introduce, for technical reasons, a cut-off function

$$\psi \in C^\infty(M \times M)$$

equal to 1 on a neighborhood of the diagonal and with support so close to the diagonal that

$$\psi(x_0, x) \neq 0, \quad v \in T^*_{x_0}\backslash 0 \implies d_x \ell(v, x) \neq 0 .$$

Then as a first approximation

$$\sigma_A(v) = A\psi(\pi(v), x)\, e^{i\ell(v,x)}\big|_{x=\pi(v)} .$$

If $A \in L^\omega_\rho$ (i.e., if A is a pseudo-differential operator of order ω and type ρ, $1-\rho$ with $\rho > \frac{1}{2}$) then $\sigma_A \in S^\omega_\rho$. Different functions ℓ and ψ give rise to σ_A differing by an element of the symbol class $S^{-\infty}$ and so the pair A, ∇ determines a unique element of $S^\omega_\rho / S^{-\infty}$ which we call the symbol of A with respect to the connection ∇. This is denoted also by σ_A. In case $M = R^n$ with ∇ the Euclidean connection σ_A is the classical symbol, mod $S^{-\infty}$.

With Taylor's formula (4.2) and the symbol defined this way one can then find the analogue of (4.1) and continue as in §2. First we introduce the operators playing the roles of $\partial^k/\partial\xi^k$ and $\partial^k/\partial x^k$.

For a function $\sigma \in C^\infty(T^*M)$ we define $D^k\sigma$ to be the k-th derivative of σ in the direction of the fibers of T^*M. Thus for $\pi(v) = x_0$ think of σ as a function on $T^*_{x_0}$ and take its k-th derivative $D^k\sigma$ evaluated at v. This is a k-linear function on $T^*_{x_0}$ and so may be identified with an element of $\otimes_k T_{x_0}$. Thus $D^k\sigma$ is a contravariant k-tensor. This is the analogue of $\partial^k\sigma/\partial\xi^k$ (although of course k is an integer in D^k and a multi-index in $\partial^k/\partial\xi^k$).

The covariant derivatives ∇^k act on $C^\infty(M)$ rather than $C^\infty(T^*M)$, and so the meaning of $\nabla^k\sigma$ for $\sigma \in C^\infty(T^*M)$ is not immediately obvious. In fact, we define it by

$$\nabla^k\sigma(v) = \nabla^k\sigma(d_x\ell(v,x))\big|_{x=\pi(v)} .$$

Although the function $\ell(v,\cdot)$ is not unique, all its derivatives are determined at $\pi(v)$, and so this defines $\nabla^k\sigma$ unambiguously as a covariant k-tensor. The mixed derivative $\nabla^k D^j \sigma$ is defined analogously by

$$\nabla^k D^j \sigma(v) = \nabla_x^k D_v^j \sigma(d_x \ell(v,x))|_{x=\pi(v)}$$

the subscripts as usual denoting the variables with respect to which the derivatives are taken. This is a tensor of type (j, k).

As coefficients in the analogue of (2.3) the unsymmetrized covariant derivatives

$$\nabla^k \ell = \nabla^k \ell(v,x)|_{x=\pi(v)}$$

will arise. This is, for each v, a tensor of type $(0,k)$. However, it is also linear in v and so $\nabla^k \ell$ itself may be thought of as a tensor of type $(1,k)$. It turns out always to be expressible as a polynomial in the torsion and curvature tensors, T and R, of the connection and their covariant derivatives. In particular we have [13, §IV]

$$\nabla^2 \ell = \frac{1}{2} T \tag{4.3}$$

and if $T = 0$

$$(\nabla^3 \ell)^p_{ijk} = \frac{1}{3}(R^p_{ijk} + R^p_{jik}), \tag{4.4}$$

where indices are used with their standard meaning. The formula is straightforward but more complicated if $T \neq 0$.

With these notations we can state the formula for the symbol of a product of pseudo-differential operators. It is

$$\sigma_{AB} = \sum_{k=0}^{\infty} \sum_{m_1 \cdots m_k \geq 2} \frac{i^{k-\Sigma_0^m p_i - \Sigma_1^k m_i}}{k! p_0! p_1! \cdots p_k! m_1! \cdots m_k!} \times \nabla^{p_1+m_1}\ell \ldots \nabla^{p_k+m_k}\ell \nabla^{p_0} D^{\Sigma m_i} \sigma_B D^{\Sigma p_i} \sigma_A . \tag{4.5}$$

The terms coming from $k = 0$ give

$$\sum_{p_0=0}^{\infty} \frac{i^{-p_0}}{p_0!} \nabla^{p_0}\sigma_B \, D^{p_0}\sigma_A$$

and in R^n with the Euclidean connection there is nothing else.

A word concerning interpretation. The differentiations $\nabla^{m_1}, \cdots, \nabla^{m_k}$ introduce Σm_i covariant indices which are to correspond to the Σm_i contravariant indices introduced by the derivative $D^{\Sigma m_i}\sigma_B$. The remaining derivatives $\nabla^{p_1}, \cdots, \nabla^{p_k}, \nabla^{p_0}$ introduce Σp_i covariant indices which correspond to the same number of contravariant indices from $D^{\Sigma p_i}\sigma_A$.

Note also that since $D^{\Sigma m_i}\sigma_B$ is a symmetric tensor any term $\nabla^{m_i}\ell$ with $p_i = 0$ may be replaced by

$$\partial^{m_i}\ell = 0$$

and so only terms with $p_1, \cdots, p_k \geq 1$ have a nonzero contribution.

Also useful, as we shall see, is a formula for change of symbol under change of connection. This is also readily derivable from the manifold analogue of (4.1) and reads

$$(4.6) \qquad \sigma'_A = \sum_{k=0}^{\infty} \sum_{m_1, \cdots, m_k \geq 2} \frac{i^{k-\Sigma m_i}}{k! m_1! \cdots m_k!} \nabla^{m_1}\ell' \cdots \nabla^{m_k}\ell' D^{\Sigma m_i}\sigma_A \, .$$

Here σ_A, σ'_A are the symbols of A with respect to connections ∇, ∇' and ℓ' is the "linear" function associated with ∇'. A special case of this is the formula, for operators on R^n, for change of symbol under change of coordinates.

Having (4.5), the manifold analogue of (2.3), we continue as in §2. Define $P_m(\sigma, \tau)$ to be the sum of those terms of the series in (4.5), with σ_A replaced by σ and σ_B by τ, corresponding to

$$\sum_0^m p_i + \sum_1^k m_i - k = m \, .$$

Then $P_m(\sigma,\tau)$ is a symbol of order

$$\text{ord}\ \sigma + \text{ord}\ \tau - (2\rho - 1)\text{m}$$

and (4.5) may be written

$$\sigma_{AB} = \sum_{m=0}^{\infty} P_m(\sigma_A, \sigma_B) .$$

Defining the Q_m and $Q_{m,k}$ by (2.4) and (2.5) we derive the formula (2.6) for manifolds.

All this can be extended to families of pseudo-differential operators T_α which locally are of type (2.1). The details of this are worked out in [13]. What happens is that with respect to any given connection the symbol of the operator T_α is of the form

$$\sigma(v/\alpha, \alpha)$$

where $\sigma(v,\alpha)$ belongs to some symbol class S_ρ^ω uniformly in α. We call this function $\sigma(v)(=\sigma(v,\alpha))$ the symbol of the family. For $\omega < -n$ the analogue of (2.2) is

$$\text{(4.7)} \qquad \text{tr}\ T_\alpha = \left(\frac{\alpha}{2\pi}\right)^n \int_{T^*M} \sigma(v)\, dv + O(\alpha^{-\infty})$$

where dv is the canonical volume element (Liouville measure) on T^*M and $O(\alpha^{-\infty})$ means $O(\alpha^{-N})$ for all N. Combining this with the analogue of (2.7) gives the asymptotic expansion of $\text{tr}\ F(T_\alpha)$ for pseudo-differential families on manifolds. Since, on manifolds,

$$Q_{1,2}(\sigma) = -i\ D\sigma\nabla\sigma$$

the generalization of (2.8) is

$$\text{tr}\ F(T_\alpha) = \left(\frac{\alpha}{2\pi}\right)^n \int_{T^*M} \left[F(\sigma) - \frac{i}{2}\alpha^{-1}F''(\sigma)\, D\sigma\nabla\sigma + \cdots\right] dv .$$

The next term is more complicated. The operators $Q_{2,k}$ involve $\nabla^3\ell$ and so, as (4.3) and (4.4) show, the curvature and torsion tensors appear in the coefficients.

With this machinery available to use in combination with the method of §3, we are now ready to find the asymptotics of

$$\log \det P(I+T_\alpha)P$$

where T_α is the operator family (2.1) on R^n and P is multiplication by the characteristic function of the bounded open set Ω with smooth boundary B. Not everything is entirely straightforward, though. The operators $+, -$ of §3 were defined in terms of the specific coordinate x_n. Thus the technique described there extends directly only to pseudo-differential families on the product manifold $B \times R$ with the half-space Ω corresponding to $B \times R^+$. Moreover, the symbols of $[S_-, S_+]$ and the other terms of the series (3.5) are readily computable in this setting only when the connection taken on $B \times R$ is the product of the connection on B (induced on B as a submanifold of R^n) with the trivial connection on R. So there are still two problems to overcome.

(1) R^n is not $B \times R$. It just looks like $B \times R$ near B.

(2) The symbol of T_α is $\sigma(x, \xi/\alpha)$ with respect to the Euclidean connection of R^n so that even if R^n were $B \times R$ the symbol would not be given with respect to the connection needed to use the method of §3.

These problems are taken care of as follows.

(1) One proves a localization theorem which says that modulo $O(\alpha^{-\infty})$ and with $\tau(v)$ denoting the symbol of the family $\log(I+T_\alpha)$,

$$\log \det P(I+T_\alpha)P - \left(\frac{\alpha}{2\pi}\right)^n \int_{T^*\Omega} \tau(v)\, dv$$

only depends on σ in any neighborhood of T^*_B. Therefore one may assume that $\sigma(x,\xi)$ vanishes for x outside a neighborhood of B in R^n which is diffeomorphic to $B \times R$.

(2) Formula (4.6) enables us to compute the symbol of T_α with respect to the product connection from the symbol with respect to the Euclidean connection.

Thus the expansion

$$\log \det P(I+T_\alpha)P = a_0\alpha^n + a_1\alpha^{n-1} + a_2\alpha^{n-2} + \cdots$$

may be thought of as having been found. The first two coefficients are in fact just given by the half-plane formulas (3.6) with z_n replaced by $z\cdot n$. The third coefficient is in general quite complicated, involving the curvature tensor of B as well as its second fundamental form; the former arises as a coefficient in (4.5), as we already saw, and the latter arises as a coefficient in (4.6). If the symbol is independent of x the formula for a_2 simplifies to the one given in the introduction. The details of all this may be found in [14].

5. *The heat expansion*

A Riemannian manifold has a natural connection, the Levi-Civita connection. With respect to it Δ, the negative of the Laplace-Beltrami operator, has symbol $\sigma(v) = |v|^2$ exactly. If the manifold is compact Δ has a discrete spectrum with eigenvalues tending to $+\infty$ and it is well known that there is an asymptotic expansion as $t \to 0+$

$$\text{(5.1)} \qquad \operatorname{tr} e^{-t\Delta} = (4\pi t)^{-n/2}(a_0 + a_1 t + a_2 t^2 + \cdots)$$

and that the coefficients a_i are expressible as polynomials in the curvature tensor and its covariant derivatives.

This may be derived from the calculus for families of pseudo-differential operators as follows [13, §VI]. Let $t = \alpha^{-2}$. The operators Δ/α^2 comprise a pseudo-differential family, i.e. locally they are of the form (2.1), with symbol $|v|^2$. (The symbol of the individual operator Δ/α^2 is $|v/\alpha|^2$.) Application of (4.7) and the manifold version of (2.7) therefore give, after replacing α by $t^{-1/2}$,

$$\operatorname{tr} e^{-t\Delta} =$$

$$(4\pi^2 t)^{-n/2} \int_{T^*M} e^{-|v|^2} \left[1 + \sum_{m=1}^{\infty} t^{m/2} \sum_{k=2}^{2m} \frac{(-1)^k}{k!} Q_{m,k}(|v|^2) \right] dv \, .$$

It is easy to see that only even m make a nonzero contribution. The $Q_{m,k}(|v|^2)$ involve the coefficients $\nabla^k \ell$ and the various $\nabla^k D^j |v|^2$, all of which are polynomials in the curvature tensor and its covariant derivatives.

The asserted form of the expansion (5.1) follows. Using this method to find the coefficients a_i is easy for $i = 0$ or 1 but quickly becomes tedious. In principle though any number are clearly computable by a mechanical procedure.

6. *Functional calculus*

Suppose A is a pseudo-differential operator on a compact manifold and F is a function such that F(A) makes sense as an operator. For example A may have order zero and F be analytic on spec A, or A may be self-adjoint and F belong to $C^\infty(R)$ and satisfy a boundedness condition. Then F(A) certainly need not be a pseudo-differential operator. It seems to be the case, though, that if the manifold version of the series (2.6) is interpretable as an element of one of the symbol classes then F(A) is pseudo-differential with that element as its complete symbol. The proof of Seeley's theorem, that F(A) is pseudo-differential if A has order zero and F is analytic on spec A [8], really uses that idea. Here are two others, which are Theorems 4.6 and 4.8 of [15]. In both, A is assumed self-adjoint and F(A) is defined using the spectral theorem.

(1) *If* $A \in L^{\omega}_{\rho}$ *with* $0 \leq \omega < \rho - \frac{1}{2}$ *and* F *satisfies*

$$F^{(k)}(x) = O(|x|^{\mu})$$

for each k *then* $F(A) \in L^{\mu\omega}_{\rho-\omega}$. (*In particular* $e^{iA} \in L^{0}_{\rho-\omega}$.)

(2) *If* $A \in L^{\omega}_{\rho}$ *with* $\omega \geq \rho - \frac{1}{2}$ *satisfies an appropriate ellipticity condition (the usual one if* $\rho = 1$) *and* F *satisfies*

$$F^{(k)}(x) = O(|x|^{\mu - \delta k})$$

for each k, *where* $\delta > 1 - \omega^{-1}(\rho - \frac{1}{2})$, *then* $F(A) \in L^{\mu\omega}_{\rho-(1-\delta)\omega}$.

In both cases, as well as in the Seeley situation, the complete symbol of F(A) is given by (2.6) suitably interpreted. Special cases of these results have been known for some time [1, 2, 9, 12].

REFERENCES

[1] Duneau, J., Fonctions d'un opérateur elliptique sur un variété compacte. J. Math. Pures et Appliqués, 56 (1977), 367-391.

[2] Guillemin, V., and Sternberg, S., On the spectra of commuting pseudo-differential operators: Recent work of Kac-Spencer, Weinstein and others, (to appear).

[3] Grenander, U., and Szegö, G., Toeplitz forms and their application. University of California Press (1958).

[4] Hadamard, J., Le probleme de Cauchy et les équations aux dérivées partielles linéaires hyperboliques, Paris (1932).

[5] Hochschild, G., The structure of Lie groups. Holden-Day (1965).

[6] Kac, M., Toeplitz matrices, translation kernels, and a related problem in probability theory. Duke Math. J. 21 (1954), 501-509.

[7] Minakshisundaram, S., and Pleijel, A., Some properties of the eigenfunctions of the Laplace operator on Riemannian manifolds. Canad. J. Math. 11 (1949), 242-256.

[8] Seeley, R. T., Complex powers of an elliptic operator. Proc. Symp. Pure Math. 10 (1968), 288-307.

[9] Strichartz, R. S., A functional calculus for elliptic pseudodifferential operators. Amer. J. Math. 94 (1972), 711-722.

[10] Szegö, G., Beiträge zur Theorie Toeplitzschen Formen I. Math. Zeitschrift 6 (1920), 167-202.

[11] ————, On certain hermitian forms associated with the Fourier series of a positive function. Festskrift Marcel Riesz, Lund (1952), 228-238.

[12] Taylor, M., Pseudodifferential operators II, (to appear).

[13] Widom, H., Families of pseudodifferential operators, in Topics in Functional Analysis (I. Gohberg and M. Kac, editors). Academic Press (1978).

[14] ———, Asymptotic expansions of determinants for families of trace class operators. Indiana U. Math. J., 27 (1978), 449-478.

[15] ———, A complete symbolic calculus for pseudodifferential operators, Bull. des Sciences Mathématiques (to appear).

Library of Congress Cataloging in Publication Data
Main entry under title:

Seminar on singularities of solutions of linear partial differential equations.

(Annals of mathematics studies; no. 91)
"Consists of the notes of a seminar held at the Institute for Advanced Study in 1977-78."
1. Differential equations, Linear—Numerical solutions—Congresses. 2. Differential equations, Partial—Numerical solutions—Congresses. I. Hörmander, Lars. II. Princeton, N.J. Institute for Advanced Study. III. Series.
QA374.S39 1979 515′.353 78-70300
ISBN 0-691-08221-9
ISBN 0-691-08213-8 pbk.